THE BIOCHEMISTRY STUDENT COMPANION

THE BIOCHEMISTRY STUDENT COMPANION

Allen J. Scism

Central Missouri State University

with an Introductory Chapter by Gale Rhodes
University of Southern Maine

PRENTICE HALL
Upper Saddle River, New Jersey 07458

Acquisitions Editor: *John Challice*
Production Editor: *Kimberly Knox*
Production Supervisor: *Joan Eurell*
Production Coordinator: *Ben Smith*
Art Director: *Joseph Sengotta*

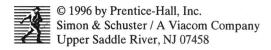
Printed in the United States of America

10 9 8 7 6 5 4 3 2 1

ISBN 0-13-449091-6

Prentice-Hall International (UK) Limited, *London*
Prentice-Hall of Australia Pty. Limited, *Sydney*
Prentice-Hall Canada, Inc., *Toronto*
Prentice-Hall Hispanoamericana, S.A., *Mexico*
Prentice-Hall of India Private Limited, *New Delhi*
Prentice-Hall of Japan, Inc., *Tokyo*
Simon & Schuster Asia Pte. Ltd., *Singapore*
Editora Prentice-Hall do Brasil, Ltda., *Rio de Janeiro*

Table of Contents

Preface

To the Student

By now, you may be wondering, "Well, I've seen all this biochemistry stuff in class and in the textbook. What exactly am I expected to get out of all this?"

Reading the text and taking good class notes are essential to succeeding in this class, of course. But no matter how well you feel you understand a particular lecture, true comprehension requires a hands-on kind of involvement. Simply put—there is no substitute for practice. In *The Biochemistry Student Companion* to accompany Horton et al. *Principles of Biochemistry 2/e*, I ask you to put pen to paper as you recall and apply what you are seeing in lectures and in the text.

To help make this book as useful as possible, I have included in the back complete solutions to all exercises. Please remember that self-discipline is an important key to effective study. Make sure you really roll up your sleeves and tackle each exercise before you consult the answers. As tempted as you may be to take a quick peek at the solution, do your best to answer the question first. If you don't, you will short-circuit your learning process.

Each chapter of *The Biochemistry Student Companion* is designed to increase and assess your comprehension of biochemistry. Section I: Key Terms lists terms within the chapter with which you should be familiar. Most of these terms are in bold type in the text and represent the vocabulary required to understand the topics within each chapter. I recommend you write out and study the definitions of all Key Terms. Section II: Exercises includes four subsections of exercises and problems. The True-False exercises require a general understanding of the material. The Short Answers exercises call for specific responses that, in many cases, require knowledge of the vocabulary of the chapter and particular facts. The Problems exercises are meant to lead you deeper into the nitty-gritty of the material, and the Additional Problems are the most challenging. In some cases, you may have to consult additional sources of information to answer these latter problems.

The Solutions section, which begins on page 99, contains full answers to all problems. I've taken great care to provide as much information in the solutions as I can; even the answers to the True-False statements include explanations. If you make a mistake, remember to take advantage of these detailed solutions to see where you went off the track.

There are some general reference tools in this book as well. In the Appendix are several tables you should find useful in your studies. Finally, the Dictionary of Biochemical Terms is a terrific reference for any biochemist—it lists over 1000 terms commonly used in the science, including many that do not appear in the text itself.

I hope *The Biochemistry Student Companion* will have a very "used" look by the time you have completed your course. May it greatly aid your understanding. Good luck, and please know that all feedback is welcome.

Acknowledgments

I would like to thank the following for their corrections and helpful suggestions: K. Gray Scrimgeour (University of Toronto), H. Robert Horton (North Carolina State University), Laurence A. Moran (University of Toronto), and Gale Rhodes (University of Southern Maine). I would also like to thank John Challice, my editor, and the folks at Prentice Hall for all their help.

Allen J. Scism
Professor of Chemistry
Central Missouri State University
ascism@cmsuvmb.cmsu.edu

An interactive computer supplement

EXPLORING MOLECULAR STRUCTURE
Version 2.0
by Kim Gernert
Duke University

This instructional software, based on Kinemages developed by David and Jane Richardson of Duke University, displays dynamic models of molecular structure. It allows students to manipulate three-dimensional depictions of biomolecules on-screen. User-friendly programming includes a built-in tutorial. On-screen text presents scripted views side-by-side with images.

Exploring Molecular Biology v2.0 contains 15 sets of exercises spanning key topics in biochemistry, and includes questions that test understanding, along with answers:

1	Protein Structure
2	Enzyme Function Part I: Specificity, Active Sites, Cofactors, and Conformations
3	Enzyme Function Part II: The Serine Proteases
4	Enzyme Function Part III: Allostery and Multidomain Proteins
5	Hemoglobin
6	Carbohydrates
7	Lipids
8	Membrane Proteins and Signal Transduction
9	Metabolism: Enzymes of the Citric Acid Cycle
10	Metabolism: Glycolysis and the Citric Acid Cycle
11	Electron Transport and Metal Centers
12	DNA and Modified DNA Structures
13	Nucleic Acid Replication, Repair, and Degradation
14	Protein-DNA Interactions
15	tRNA and Synthetases

available in these formats: Macintosh (ISBN 0-13-458472-4)
Windows (ISBN 0-13-458464-3)

System requirements:

Macintosh: 1 MB RAM. Color monitor is recommended but not required.

Windows: 1 MB RAM and Windows 3.1 or later release. Color monitor is recommended but not required.

Order through your college bookstore or from

PRENTICE HALL
Order Department
200 Old Tappan Road
Old Tappan, NJ 07675
1-800-947-7700

1

Prelude to Biochemistry: A Review of Important Concepts from Organic Chemistry

To the Biochemistry Student

A useful definition of *biochemistry* is "the tools of chemistry and physics applied to the problems of biology." You can interpret this definition in two ways. First, much of the progress in biochemistry is driven by the experimental tools of chemistry and physics, including structural analysis, kinetics, thermodynamics, radioisotope labeling, spectroscopy, crystallography, and molecular modeling. Understanding biochemistry as a dynamic, evolving field means seeing how currently accepted models have been devised by interpreting data obtained using chemical tools. Second, and of the most immediate importance as you begin your study of biochemistry, you will need to use your own tools of chemical understanding, which you developed in general and organic chemistry. This may mean that you will need to review material from previous courses, especially organic chemistry. This chapter provides a brief review of concepts from organic chemistry that you are most likely to need in biochemistry. I hope this review is adequate to get you started, but if any subject I raise here is unfamiliar, look it up in the index of your organic chemistry text to find further information.

NOTE: Terms in *italics* are *biochemical terms*, and they may be entirely new to you, but I will explain them where they first appear. Terms in **bold** are **review terms**, and they should be somewhat familiar. Each review term will appear in one or more biochemical examples designed to help you recall the term and its meaning.

Overview

To begin this review, study Figure 1, which shows a sequence of ten chemical reactions that constitute the *metabolic pathway* called *glycolysis*. This sequence of reactions occurs in almost all living organisms. One of its main functions is to derive energy, in the form of the chemically active substance *ATP*, from the nutrient glucose. Each compound in the pathway is a product of the previous reaction and a reactant in the succeeding reaction, and each is therefore called a *metabolic intermediate*. Other reactants and products are shown entering and leaving the pathway by arrows above or below the main reaction arrows. You will study this pathway in more detail later. For now, examining it will serve to remind you of, and to illustrate, many areas of basic organic chemistry as they apply to biochemistry. For convenience, the reactions of the pathway are numbered, and each compound is labeled with its full and abbreviated name. I will use only the abbreviated names in the text. Don't worry about remembering the names and structures of all these compounds. Focus instead on the **review terms** and try to understand how they apply in these new surroundings.

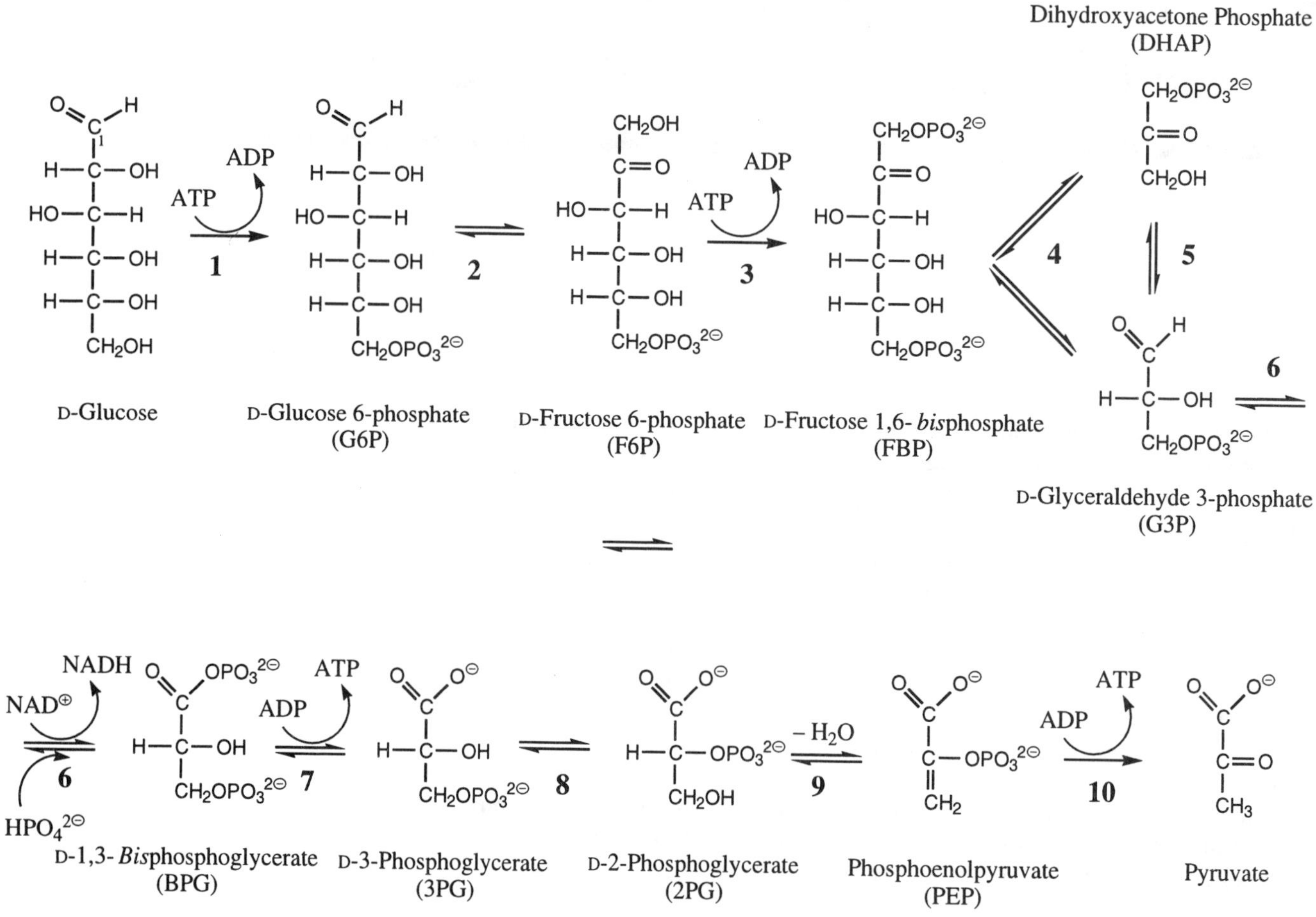

Dihydroxyacetone Phosphate
(DHAP)

$CH_2OPO_3{}^{2\ominus}$
$C=O$
CH_2OH

O=C—H
H—C—OH
HO—C—H
H—C—OH
H—C—OH
CH_2OH
D-Glucose

ADP
ATP
1

O=C—H
H—C—OH
HO—C—H
H—C—OH
H—C—OH
$CH_2OPO_3{}^{2\ominus}$
D-Glucose 6-phosphate
(G6P)

2

CH_2OH
$C=O$
HO—C—H
H—C—OH
H—C—OH
$CH_2OPO_3{}^{2\ominus}$
D-Fructose 6-phosphate
(F6P)

ADP
ATP
3

$CH_2OPO_3{}^{2\ominus}$
$C=O$
HO—C—H
H—C—OH
H—C—OH
$CH_2OPO_3{}^{2\ominus}$
D-Fructose 1,6-*bis*phosphate
(FBP)

4
5

O=C—H
H—C—OH
$CH_2OPO_3{}^{2\ominus}$
D-Glyceraldehyde 3-phosphate
(G3P)

6

NADH
NAD$^{\oplus}$
6
HPO$_4{}^{2\ominus}$
O $OPO_3{}^{2\ominus}$
\C/
H—C—OH
$CH_2OPO_3{}^{2\ominus}$
D-1,3-*Bis*phosphoglycerate
(BPG)

ATP
ADP
7
O $O^{\ominus}$
\C/
H—C—OH
$CH_2OPO_3{}^{2\ominus}$
D-3-Phosphoglycerate
(3PG)

8
O $O^{\ominus}$
\C/
H—C—$OPO_3{}^{2\ominus}$
CH_2OH
D-2-Phosphoglycerate
(2PG)

− H₂O
9
O $O^{\ominus}$
\C/
C—$OPO_3{}^{2\ominus}$
‖
CH_2
Phosphoenolpyruvate
(PEP)

ATP
ADP
10
O $O^{\ominus}$
\C/
$C=O$
CH_3
Pyruvate

Figure 1. The Ten Reactions of Glycolysis

Before I describe the pathway in detail, here's a sampling of the concepts I will consider. I will draw your attention first to the <u>structures</u> of the intermediates, and second to the <u>reactions</u> depicted. First, look at structures. Try giving **systematic names** to a few of them, using *oxyphosphoryl* as the **substituent** name of the $OPO_3{}^{2\ominus}$ group. Your names will not be the same as the commonly used names; for example, the oxyphosphoryl group is usually called a **phosphate** group, as in glucose 6-phosphate, the product of the first reaction. Among the intermediates, can you find pairs of **structural isomers**? What **functional groups** can you find and name? What functional groups are stabilized by **resonance**? Do you remember the meaning of resonance, and how to draw the most important structures that contribute to a **resonance**

hybrid? Which intermediates are **chiral compounds** and how many **stereoisomers** are possible for each structure? Biochemists use D- and L- to designate the **configurations** of chiral compounds. Recall that each **chiral center** can also be assigned the **(R)**- or **(S)**-configuration, according to the system of Cahn, Ingold, and Prelog. Which molecules might exist in equilibrium with forms other than those shown? For example, which might exist in two or more **tautomeric forms**, and which contain two functional groups that might combine with each other in an **intramolecular reaction** to produce an alternative form?

Second, examine each reaction. Try to discern what structural change converts the reactant to the product, and try to classify each change as **addition**, **elimination**, **substitution**, or combinations of these simple **reaction types**. Can you write **mechanisms** for these reactions, assuming some simple type of **catalysis**, like **specific-acid catalysis** (catalysis by protons), and depicting with **curved arrows** the changing allegiances of electrons as bonds break and form? (This powerful form of writing reaction mechanisms is sometimes called **electron pushing**.) Do you see any **oxidation** or **reduction** reactions? One way to detect such changes is to compute the average **oxidation number** of carbon for each structure, assuming that the oxidation number of hydrogen is +1 and that of oxygen is –2. (To simplify detection of oxidation-number changes in carbon, you can ignore complex groups that do not change during the reaction.)

Now you have an idea of some of the areas of organic chemistry that are essential to understanding a metabolic pathway, and some of the concepts that you need to review and use in your study of biochemistry. Let's begin our detailed review by looking at the pathway more closely.

The Structures

The first structure in Figure 1 could be called 2,3,4,5,6-pentahydroxy-hexanal, but fortunately it has a common name, **D-glucose**. In fact, many biological compounds have common names that were adopted before their structures were known. This prevalence of common or trivial names is in part a blessing, because the names are often much shorter than systematic names, and in part a curse, because you cannot figure out the structures from the names. You must therefore simply memorize many new names. Looking at the structure of glucose, you can probably deduce that this compound is water-soluble. Organic compounds having at least one **hydroxyl** (—OH) or other **hydrogen-bonding** group per three or four carbons are usually appreciably soluble in water. Looking over all the intermediates in glycolysis, you can see that all would be very soluble in water. Not all biomolecules are water-soluble, however. *Fats and oils*, which are **esters** of *fatty acids* such as *palmitic acid* ($CH_3(CH_2)_{14}COOH$), are quite insoluble in water because they are mostly **nonpolar**. They are quite soluble in organic solvents like hexane. Many water-insoluble biomolecules belong to the class called *lipids*. Some are involved in forming *membranes*, which provide a barrier that makes the cell and its internal compartments impermeable to many water-soluble substances.

The sugar glucose is a **monosaccharide**, a simple **carbohydrate**. Carbohydrates, one of the four main types or classes of biomolecules, include sugars and starches (the other biomolecules are *lipids*, *proteins*, and *nucleic acids*). Most monosaccharides are polyhydroxycarbonyl compounds, either **aldehydes** like glucose, or **ketones** like fructose. Notice that glucose has four **stereocenters** or **chiral centers**, at carbons 2, 3, 4, and 5 (the aldehyde carbon is C-1). These carbons form covalent bonds (which means that they share electron pairs) through ***sp³*-hybridized orbitals**, so they form four bonds with

tetrahedral geometry. (C-1, in contrast, is *sp²-hybridized* and **trigonal planar**.) Because two **configurations** are possible at each stereocenter, there are 2^4 or 16 different **stereoisomers** of 2,3,4,5,6-pentahydroxyhexanal. Only the full mirror image, or the **enantiomer**, of D-glucose, shares its name. The enantiomer shown is D-glucose, designated D- because of its structural kinship with the simpler chiral compound D-*glyceraldehyde* (the 3-phosphate derivative of D-glyceraldehyde, called D-glyceraldehye 3-phosphate, is one of the products of Reaction 4 in Figure 1). The enantiomer of D-glucose is called L-glucose. The other stereoisomers of glucose, which are its **diastereoisomers** or **diastereomers**, are chemically distinct from D- and L-glucose, and have their own common names.

The drawings of glucose and the other intermediates in Figure 1 are called **Fischer projections**. At each chiral center, the groups attached by horizontal bonds in Fischer projections are understood to be projected outward toward the reader, while the groups attached by vertical bonds are pointing back into the page. Because the chiral carbons are sp^3 hybridized, the angle formed by the chiral carbon and any pair of attached atoms is approximately 109.5 degrees.

Because glucose is an aldehyde with a hydrogen atom on its α-carbon (C-2), in aqueous solutions it exists in equilibrium with a **tautomeric enol** form. The equilibrium reaction is shown in Figure 2.

Carbons next to carbonyls are weakly acidic. If a base abstracts a pro-

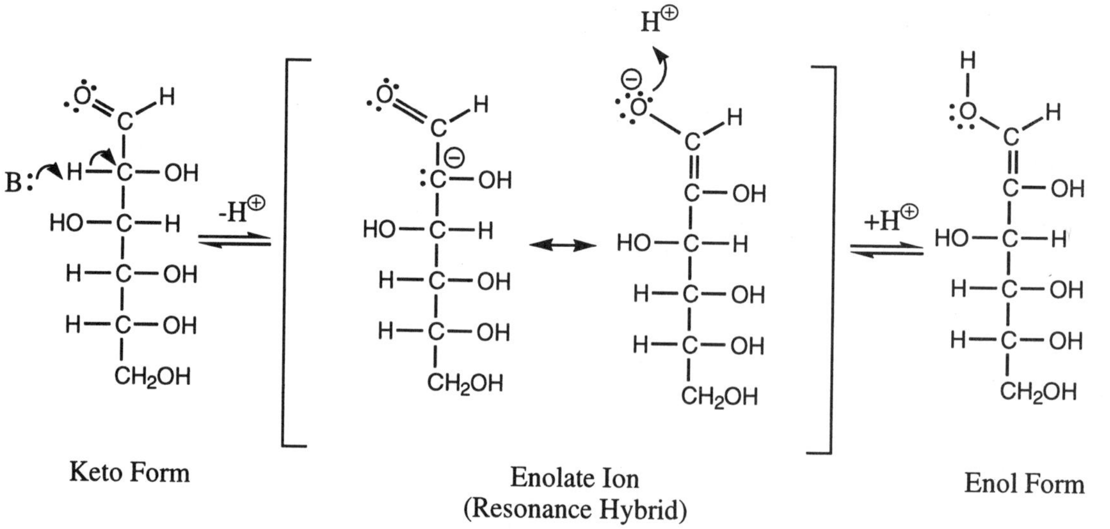

Figure 2. Tautomerism in Glucose

ton from the α-carbon of the aldehyde (which is called the **keto** tautomer), a resonance-stabilized **enolate ion** (shown in square brackets) results. The enolate may be reprotonated (almost certainly by a different proton from the solvent) at C-2 or at O-1 (the oxygen of C-1). Protonation at O-1 (shown) produces the enol form of glucose. This process is reversible, and because the reactant, glucose, is less acidic than the enol, it is more stable and predominates in the tautomeric equilibrium. The fraction of enol-glucose at equilibrium is so small that we always write glucose as shown in Figure 1, but tautomerism is important in explaining both spectroscopic and chemical properties of aldehydes and ketones, as I will show later.

Now examine the structures shown within the square brackets in Figure 2. These structures, taken together, depict the enolate ion as a **resonance hybrid** of the structures shown, which are called **resonance contributors**. Organic chemists depict molecules or ions as resonance hybrids when one Lewis diagram is inadequate to depict them fully. In this enolate ion, the negative charge is distributed between C-2 and the oxygen of C-1, so neither of the two drawings alone fully depicts the properties of the ion. This terminology does <u>not</u> imply that an enolate is an equilibrium mixture of the resonance contributors. It is better to say that the enolate ion is a <u>composite</u> of the contributors, which means that it has some of the properties of both.

A telling analogy from an organic text of the late sixties helped me understand the meaning of resonance. It goes like this:

> A rhinoceros is a resonance hybrid of a dragon and a unicorn. This does not imply that the rhino spends part of its time as a dragon and part of its time as a unicorn. Instead, it means that the rhino has some dragon attributes (like a tough, thick coat), and some unicorn attributes (like its horn). In addition the rhino, like the resonance hybrid, is a real animal, but the dragon and unicorn, like the contributors, are fictional.

So actually, the resonance contributors do not exist at all. They are fictional characters whose structures suggest the properties we want to depict for the real enolate ion. We view the hybrid as being most like the contributor or contributors we judge to be the most stable, based on criteria such as charge distribution and number of covalent bonds. Contributors with less separation of opposite charge and more shared electrons are more stable. And the more equivalent forms we can write, the greater is the stabilization by resonance. (While I think of the contributors as useful fictions, I leave for philosophers the question of whether molecules and ions themselves are real. Like the vivid characters in good fiction, they certainly help us to see the real world more clearly.)

Looking again at the enolate ion, you should conclude that, if the two contributors were distinct substances, the second contributor would be more stable because the negative charge resides on oxygen, which is more electronegative than carbon. Chemists assume, therefore, that the enolate ion is more like the more stable contributor, so you would expect the enolate ion to protonate faster on oxygen because a greater portion of the negative charge resides there. If so, why is the keto rather than the enol form favored at equilibrium? It must be that the lifetime of the enol form is shorter than that of the keto form, or to put it another way, the enol form is more acidic. The faster protonation of the enolate ion at oxygen is a **kinetic** effect, while the predominance of the keto form at equilibrium is a **thermodynamic** effect. In chemical reactions that can produce two alternative products, the one that forms faster is called the **kinetic product**, while the more stable one is called the **thermodynamic product**. In this example, the enol form is the kinetic product of protonation, but the keto form is the thermodynamic product. Given enough time, the keto form will predominate. Chemists can often control reaction conditions to obtain a desired product. For instance, short reaction times favor kinetic products, while long reaction times at high temperatures favor the thermodynamic product by providing plenty of opportunity for the reaction to come to equilibrium.

Finally, notice the three curved arrows in Figure 2. These arrows depict the movement of electrons as bonds break and form. In the first step of the mechanism shown, an unspecified base removes a proton (hydrogen ion) from glucose. The curved arrow shows that the bonding electron pair in the C–H bond becomes an unshared pair on carbon, leaving the proton with no electrons. In the second step, the curved arrow shows that an unshared electron pair on oxygen becomes a bonding pair between oxygen and a proton. Notice that the arrows depict the movement of *electrons*, not nuclei — for example, no arrow follows the lost proton. In writing reaction mechanisms, curved arrows are used only to depict electron movement.

Now focus on the various functional groups of the intermediates in Figure 1. Glucose, with its **carbonyl** ($>$C=O) group at C-1, is an **aldehyde**. It also possesses **hydroxyl** (—OH) groups, the functional groups found in **alcohols**. The product of Reaction 2 in Figure 1, fructose 6-phosphate, or F6P, is a **ketone**, because the carbonyl carbon is attached to two carbons. Like glucose, it is also an alcohol. Joined to O-6 of fructose is a **phosphate** (—OPO$_3^{2\ominus}$) group. F6P is thus also an **ester**, which is the product of reaction between an alcohol and an acid. You are most familiar with esters of **carboxylic acids**, which contain **carboxyl** groups (—COOH). F6P is the product of reaction between an alcohol (the C-6 hydroxyl of fructose) and phosphoric acid. Figure 3a compares a carboxylate ester with a phosphate ester. Notice the similarity in bonding in the ester linkages (in boxes).

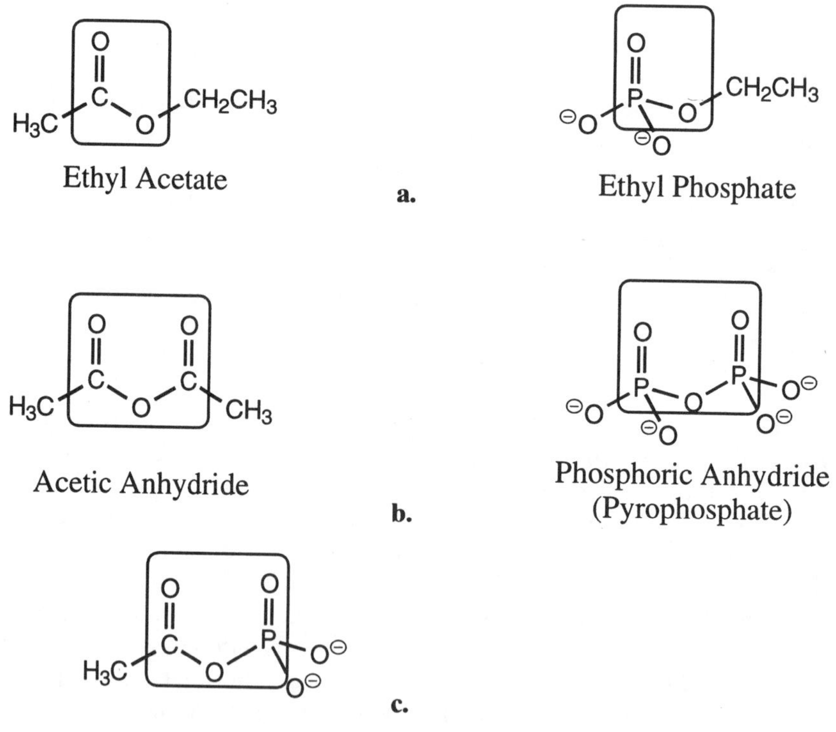

Ethyl Acetate   **a.**   Ethyl Phosphate

Acetic Anhydride

b.

Phosphoric Anhydride
(Pyrophosphate)

c.

Acetyl Phosphate, a Mixed Anhydride

Figure 3. Analogous Carbon and Phosphorus Compounds. **a.** Carboxylate and Phosphate Esters **b.** Carboxylate and Phosphate Anhydrides **c.** A Mixed Anhydride

The product of Reaction 6 (Figure 1), BPG, is an **anhydride**. As with esters, you are most familiar with anhydrides of carboxylic acids. 1,3-*Bis*phosphoglyceric acid (BPG) is a **mixed anhydride** of phosphoric acid and a carboxylic acid, 3-phosphoglyceric acid. Figures 3b and 3c compare carboxylic, mixed carboxylic-phosphoric, and phosphoric anhydrides (called **phosphoan-**

hydrides) Notice the similarity in bonding in the anhydride linkages (in boxes). **Hydrolysis** (literally, cleavage by water) of anhydrides gives acids.

Reaction 7 produces a carboxylic acid (3PG) by transfer of the phosphate from BPG to *adenosine 5′-diphosphate* (*ADP*). The carboxyl group of 3PG is shown in Figure 1 in its unprotonated, or **carboxylate**, form, because this form would predominate at pH values common to the cellular and intracellular fluids of living organisms (pH 7 to 8.5). The pK_a values for simple carboxylic acids are well below 7. When the pH of the solution is numerically the same as the pK_a of an acid, the acid is 50% ionized; that is, the protonated and unprotonated forms are present in equal quantities. At higher pH (lower concentration of hydrogen ions), the unprotonated form predominates by a factor of 10 for each unit of difference between the pH and the pK_a. For example, if the pK_a of an acid is 4.0, then at pH 7.0, the unprotonated groups (carboxylates) outnumber the protonated groups (carboxyls) by 1000 to 1. At pH 4.0, carboxylates and carboxyls are present in equal quantities.

The carboxylate group provides another example of resonance stabilization, as shown for acetate ion in Figure 4. Curved arrows on the first contributor show that you can use electron pushing to help you write resonance

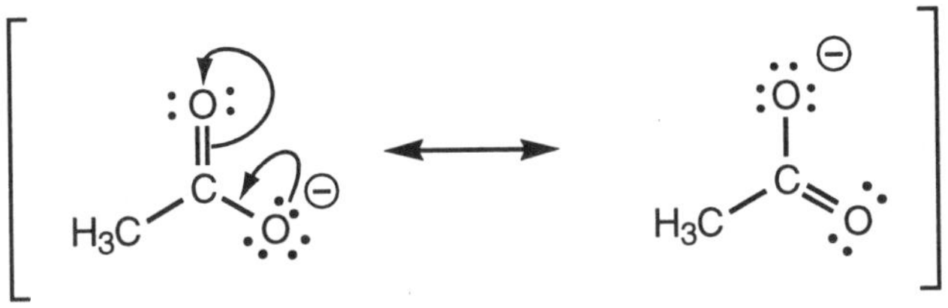

Figure 4. Resonance Stabilization of Acetate Ion

contributors. See if you can use curved arrows to generate the second contributor for the enolate ion in Figure 2.

The two resonance forms of the carboxylate ion are **equivalent**; that is, we judge them to be of equal energy, because bonding and charge distribution are identical in the two forms. For this reason, we expect resonance stabilization to be greater in the carboxylate ion than in the enolate ion, in which the resonance forms are not equivalent (one carries negative charge on carbon, the other on oxygen). This greater resonance stabilization of the carboxylate partially explains the much greater acidity of carboxyl groups (with typical pK_a values between 2 and 6) in comparison to carbon acids like ketones or aldehydes having α-hydrogens (with typical pK_a values greater than 20).

Glucose provides one more opportunity to review a class of organic compounds that is important in carbohydrate chemistry. Glucose contains the aldehyde carbonyl as well as the hydroxyl groups found in alcohols. Alcohols react with aldehydes (and ketones) to form, first, **hemiacetals**, R-CH(OH)(OR′), and then **acetals**, R-CH(OR′)$_2$, (with ketones the products are **hemiketals** and **ketals**, but in modern nomenclature, *acetal* is the preferred term for both classes). Figure 5a shows a simple example of this reaction between benzaldehyde and methanol.

Again, note the curved arrows that show bond breakage and bond formation. When you look at a detailed mechanism, notice that the arrows on the reactants in each step provide a set of instructions for drawing the products of that step. To draw the products, simply redraw the reactants with the bonds moved as indicated by the curved arrows. Thinking about electron pushing in this way will help you to read mechanisms as well as to write reasonable and clear mechanisms. Often, trying to write a mechanism that explains how reactants becomes products helps you to understand the details and consequences of chemical change, as well as the roles of the various agents (like hydrogen ions) that promote reactions.

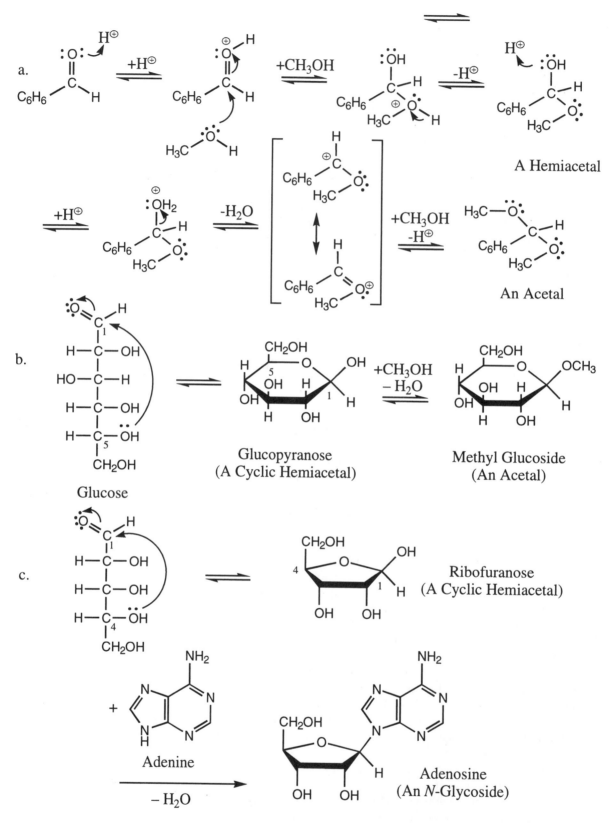

Figure 5. Hemiacetals and Acetals **a.** Reaction of Benzaldehyde with Methanol **b.** Formation of the Cyclic Hemiacetal of Glucose and Methyl Glucoside, an Acetal **c.** Formation of the Cyclic Hemiacetal of Ribose and Adenosine, an *N*-glycoside

The presence of both the aldehyde carbonyl and hydroxyl groups in glucose makes possible intramolecular hemiacetal formation. Intramolecular reactions are particularly likely when they produce five- or six-membered rings. It is therefore not surprising that glucose in water forms a cyclic hemiacetal. The first step in Figure 5b shows the formation of this cyclic form, called glucopyranose, from the open-chain form of glucose. The hydroxyl of C-5 reacts with the C-1 carbonyl to form a six-membered ring. See if you can write a mechanism for this process, using the mechanism in Figure 5a as a guide. Assume acid catalysis.

When you encounter glycolysis later in your course, you will probably find that glucose is drawn in this glucopyranose form, which predominates in solution. The hemiacetal hydroxyl, on C-1 in the case of glucopyranose, is quite reactive in substitution reactions with other alcohols or amines. Other sugar molecules can join by way of their hydroxyls with C-1 of glucose to form **glucosides** such as complex sugars and starches, which are acetals of glucose. The general term for an acetal of a monosaccharide is **glycoside**. The second step in Figure 5b shows the formation of a methyl glycoside from glucose. Some sugars, like ribose, react with amines to form *N*-**glycosides**, as shown in Figure 5c You will find examples of *N*-glycosides in DNA and RNA. Try writing a mechanism for the reaction in Figure 5c. Assume acid catalysis.

Amines and **thiols** are among other compounds that are important in biochemistry. Amines, whether **primary** (RNH_2), **secondary** (R_2NH), or **tertiary** (R_3N), are organic bases. The pK_a values for the conjugate acids of most amines (for instance, $RNH_3^\oplus$, a **primary ammonium ion**) are higher than 9.0, so protonated forms of amines predominate under physiological conditions. Simple **thiols** (RSH) have pK_a values around 9.0, so a significant amount of the thiolate form ($RS^\ominus$) can exist at neutral pH, although RSH predominates. Both thiols and thiolates are powerful **nucleophiles**. An important reaction of thiols is oxidation to **disulfides** (RSSR). This reaction occurs spontaneously in the presence of the oxygen in air. Disulfides stabilize the structures of some proteins.

The Reactions

Having used some of the intermediates of glycolysis to review many aspects of structural organic chemistry, let's go through the pathway step by step, looking at reactions. In the cell, each reaction is catalyzed by a different *enzyme*. Enzymes are *proteins* that are highly specific catalysts that greatly enhance the rates of metabolic reactions by providing reaction pathways with lower energy barriers than those of uncatalyzed reactions. Because enzyme action is complex, and is an important subject for later study in your biochemistry course, I will look at each reaction as if it were promoted by a simple catalyst like $H^\oplus$ or $OH^\ominus$, just as I did in the the cyclization of glucose. Thus the mechanisms I write are not accurate depictions of these reactions as they occur in the cell, with enzymatic catalysis, but instead are designed to remind you of the general nature of some common organic reactions.

Reaction 1 in Figure 1 is the conversion of glucose to G6P. Specifically, the reaction produces a **phosphoester** from an alcohol (the primary alcohol at C-6 of glucose) and the phosphoanhydride ATP. The reaction is shown in more detail in Figure 6b. Don't let the complex structure of ATP scare you. In almost all reactions involving ATP, all the chemical action occurs at the relatively simple phosphate anhydride linkages. The rest of the molecule is a handle by which the appropriate enzymes recognize the molecule.

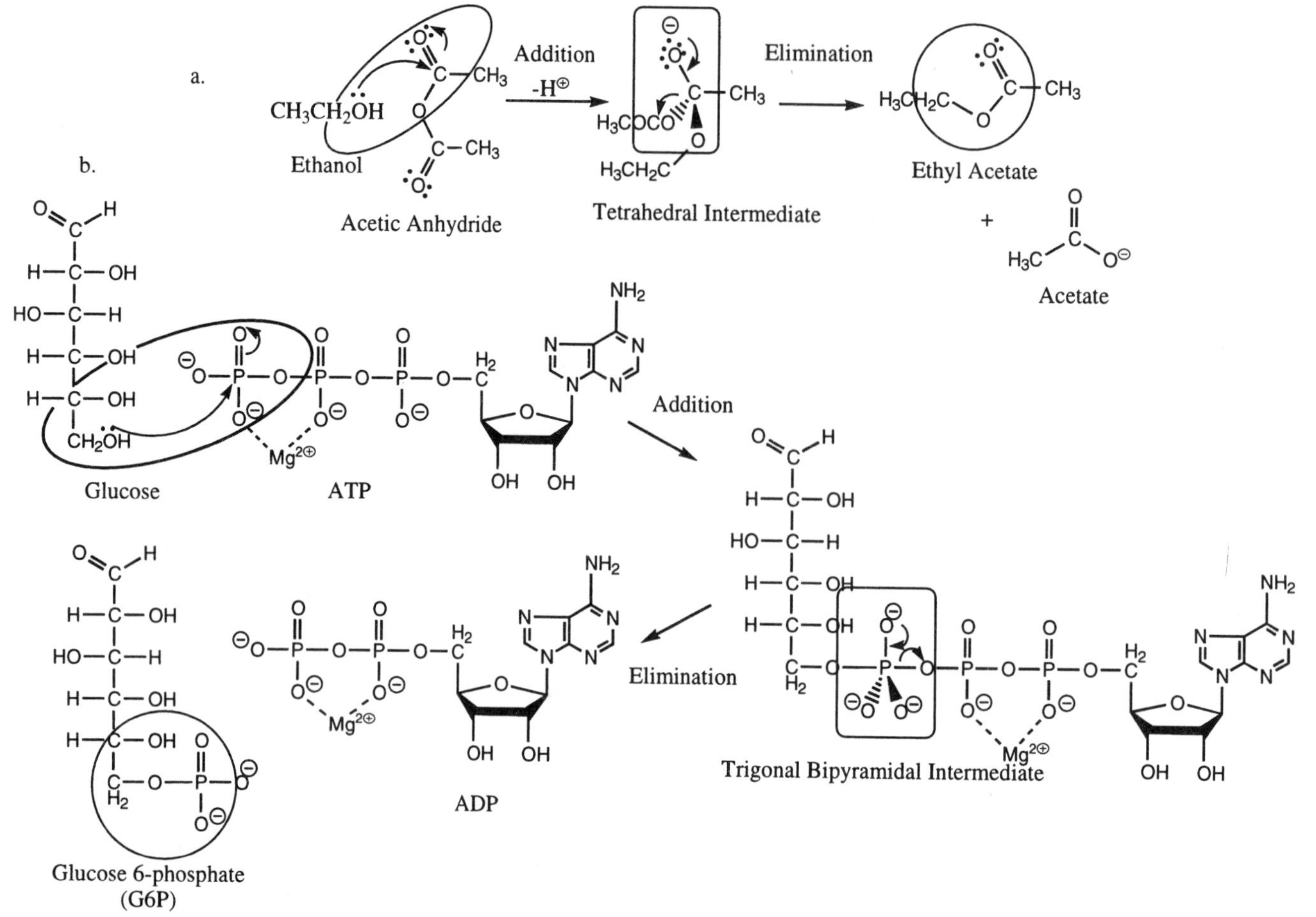

Figure 6. Esterifications **a.** Carboxylate Ester Formation (Ethyl Acetate) **b.** Phosphate Ester Formation (G6P)

Formation of G6P is analogous to formation of a simple ester from a carboxylic anhydride and an alcohol, as illustrated in Figure 6a. Compare this reaction with the conversion of glucose to G6P in Figure 6b. In the two mechanisms, analogous linkages are enclosed in similar figures (compare ovals with ovals, rectangles with rectangles, and so forth). Notice that the first step in each reaction is **nucleophilic addition** to a double bond. Addition to the carboxylate carbon gives a tetrahedral intermediate, while addition to the phosphate phosphorus gives a pentavalent **trigonal bipyramidal** intermediate. The final step, which reestablishes the double bond, is **elimination**. The two steps together constitute a **substitution** reaction. Derivatives of carboxylic and phosphoric acids readily undergo this type of substitution.

The reactions of Figure 6 occur spontaneously under standard conditions (that is, their equilibrium constants are greater than 1.0) because anhydrides like ATP are thermodynamically less stable than esters like G6P. Thus the reactants (anhydride plus alcohol) are less stable than the products (ester plus acid), and products predominate at equilibrium. In general, anhydrides are the least stable compounds derived from carboxylic acids. Roughly speaking, the order of thermodynamic reactivity for biologically important carboxyl derivatives and analogous phosphate compounds is as follows:

$$\underset{\text{anhydride}}{\text{carboxyl}} \geq \underset{\text{anhydride}}{\underset{\text{(carboxyl-phosphate)}}{\text{mixed}}} > \text{phosphoanhydride} \geq \underset{\text{(RCOSR}^-)}{\text{thioester}} > \text{ester} > \underset{\text{(RCONH}_2)}{\text{amide}} > \text{acid}$$

This order largely reflects the stability of the group that leave the carbonyl carbon during hydrolysis. The most reactive carboxyl derivatives release the least basic leaving groups. Further, this order reflects <u>thermodynamic</u> stability, which is related to equilibrium constants for reactions, as opposed to <u>kinetic</u> stability, which is related to rates of reactions. You can use this order of reactivity to decide whether a reaction has a favorable equilibrium constant ($K > 1$). In general, the formation of any compound type from one before it in the reactivity series is favorable. For example, because esters are higher in the series than amides, treatment of an ester with an amine to form an amide is a favorable process. Referring again to Reaction 1, the formation of the phosphate ester G6P from the phosphanhydride ATP is a favorable process.

All reactions involving ATP require the presence of a divalent metal ion, most commonly $Mg^{2\oplus}$. $Mg^{2\oplus}$ forms complexes with phosphate oxygen atoms on ATP. Several different types of complexes are possible, and representative structures are shown in Figure 6b. $Mg^{2\oplus}$ partially neutralizes negative charges on the oxygen atoms, and thus makes ATP more susceptible to nucleophilic attack. As a result, $Mg^{2\oplus}$—ATP is a better **electrophile** than free ATP. This role is one of many for metal ions in biochemistry. Ions are essential components of some enzymes, where they may act as **Lewis acids** (electron acceptors) or as **oxidation-reduction (redox)** agents. Some biochemistry texts represent bonding to metal ions with dotted lines, implying electrostatic attraction, while others represent it with solid lines, implying covalent bonding. Inorganic chemists use several bonding theories to explain bonding in metal-ion complexes, but it is usually safe to interpret interactions between metal ions and electron-donating species, called **ligands**, as being electrostatic in nature, with ligands donating electrons into empty orbitals of the metal ion. You can judge the hybridization of the metal ion by the geometry of the complex, the most common cases being **octahedral** (d^2sp^3, 6 ligands around the metal ion, or a **coordination number** of 6), **trigonal bipyramidal** (dsp^3, 5 ligands), and **tetrahedral** (sp^3, 4 ligands).

Reaction 2 in Figure 1, the formation of F6P from G6P, produces a ketone from an aldehyde. Look again at the enol form of glucose in Figure 2. You can see that it is in fact an **enediol**, because both carbons of the double bond carry hydroxyls. This means that a third tautomer is possible, with a carbonyl at C-2. This tautomer is fructose. It is thought that the conversion of G6P to F6P proceeds by way of this enediol intermediate. Try to write a four-step mechanism for this process (the three intermediates are 1. enolate ion; 2. enediol; 3. enolate ion). Assume acid catalysis.

In Reaction 3 of Figure 1, F6P accepts a second phosphate group from ATP, producing FBP, in a reaction similar to Reaction 1. In Reaction 4, FBP is split into two 3-carbon sugars by a reaction whose reverse is a familiar carbon-carbon bond-forming reaction, the **aldol condensation**. Figure 7 compares a simple reverse aldol reaction (a) with reaction 4 (b). In Figure 7a, the reverse-aldol cleavage of a β-hydroxy aldehyde is catalyzed by hydroxide ion, an example of **specific base catalysis**. In Figure 7b, the reaction is catalyzed by an unspecified base, which could be a basic functional group on the enzyme that catalyzes this reaction. This is an example of **general base catalysis**, which simply means catalysis by a base other than hydroxide ion.

A reverse aldol reaction requires a β-hydroxy aldehyde or ketone. Because most monosaccharides contain hydroxyl groups at their β positions (two carbons away from the carbonyl, or at C-3 in FBP), reverse aldol reactions are common in carbohydrate chemistry.

Of the two products from reaction 4, only G3P can take part in Reaction 6. DHAP is not wasted, however, because it is converted to a second molecule of G3P by Reaction 5, which is very similar to Reaction 2. Write a mechanism for the acid-catalyzed formation of G3P from DHAP, by analogy to the mechanism of Reaction 2.

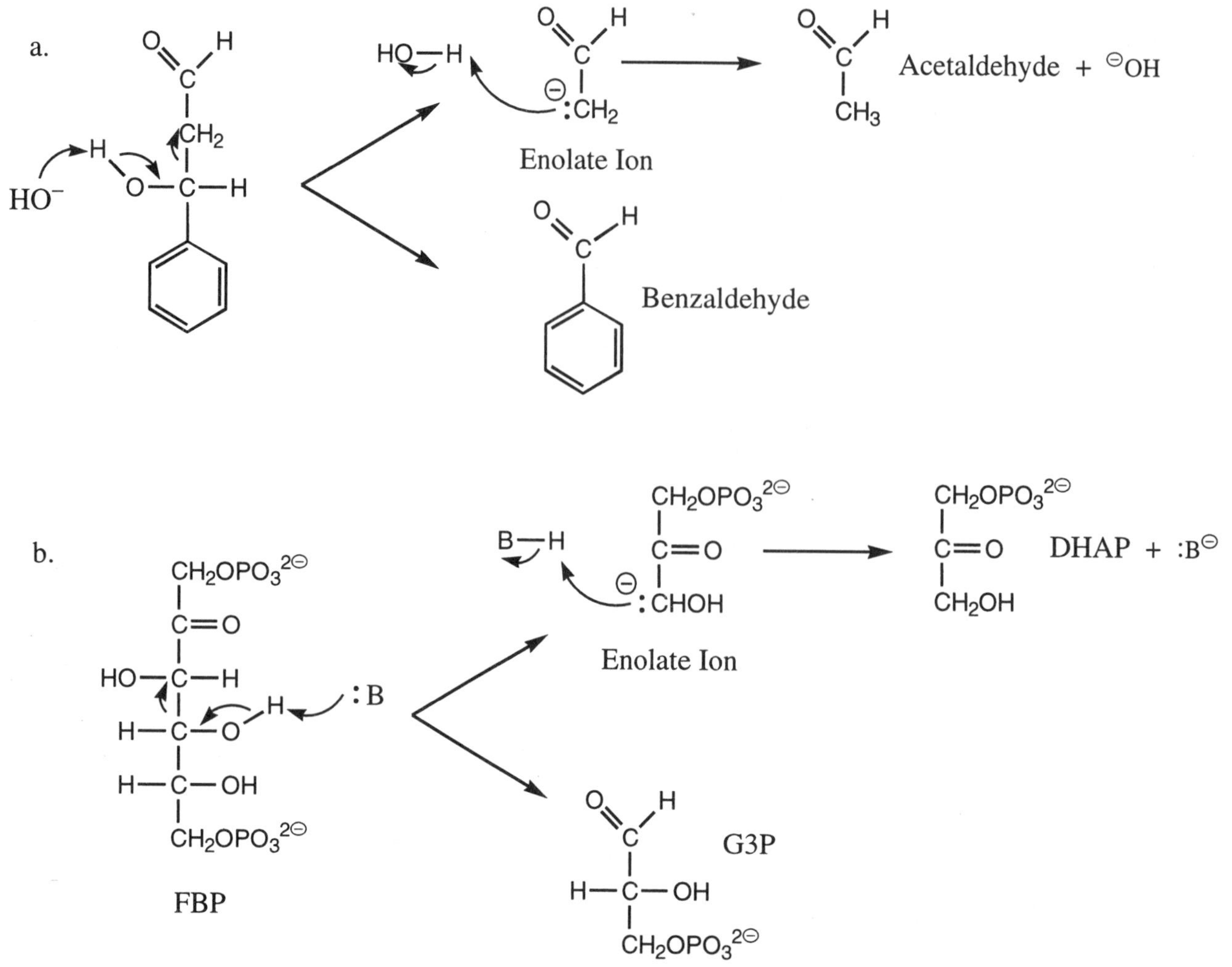

Figure 7. a. Reverse Aldol Reaction. **b.** Conversion of FBP to DHAP and G3P.

Both molecules of G3P continue through the pathway, so two molecules of pyruvate are formed from each molecule of glucose that enters glycolysis. This conclusion can be tested by **radioisotope labeling** experiments, in which the fate of specific atoms can be followed by replacing them with radioactive isotopes in a small fraction of the molecules. For example, if FBP containing the radioactive isotope ^{14}C at position C-4 is added to a cell extract undergoing active glycolysis, all of the radioactivity is recovered in C-1 of pyruvate. If the experiment is repeated using ^{14}C-3 FBP, the radioactivity is again found in C-1 of pyruvate, demonstrating that both DHAP and G3P are converted to pyruvate by the enzymes of glycolysis. Biochemists have used radioisotopic labeling extensively in establishing the mechanisms of reactions and the order of intermediates in metabolic pathways.

Reaction 6 converts G3P and a phosphate ion to 1,3-BPG. To help recognize what type of reaction this is, calculate the **oxidation number** of C-1 in the reactant and product. Here is a two-step procedure for computing oxidation numbers within a complete structural formula. First, assign all electrons to atoms as follows: assign unshared electrons to the atom on which they reside, and bonding electrons to the more electronegative of the two bonded atoms; if a bond joins two atoms of the same element, assign one bonding electron to each. Second, compute the oxidation number for any atom by subtracting the number of assigned electrons from the group number of that element in the Periodic Table. For C-1 of G3P, the oxidation number is [(group number for carbon) − (assigned electrons)] = 4 − 3 = +1. The oxidation number of C-1 in 1,3-BPG is 4 − 1 = +3. Reaction 6 thus increases the oxidation number of C-1, which means that this reaction is an **oxidation**.

Oxidation is always accompanied by **reduction**. The substance that becomes reduced during the reaction is called the **oxidizing agent**. The oxidizing agent in Reaction 6 is $NAD^{\oplus}$, and the reduced product is NADH. $NAD^{\oplus}$ is derived from nicotinamide, a *vitamin*. The oxidized and reduced forms of the nicotinamide ring, which are the reactive parts of the structures of $NAD^{\oplus}$ and NADH, are shown in Figure 8 (the R group, like most of the ATP molecule, is a large, complex "handle" for enzymes). C-4 of $NAD^{\oplus}$ is thought to accept a hydride ion ($H^{\ominus}$) from C-1 of G3P in this reaction. As an exercise in computing oxidation numbers, calculate oxidation numbers for C-4 in $NAD^{\oplus}$ and in NADH, and thus convince yourself that this carbon is reduced (its oxidation number <u>decreases</u>) during Reaction 6.

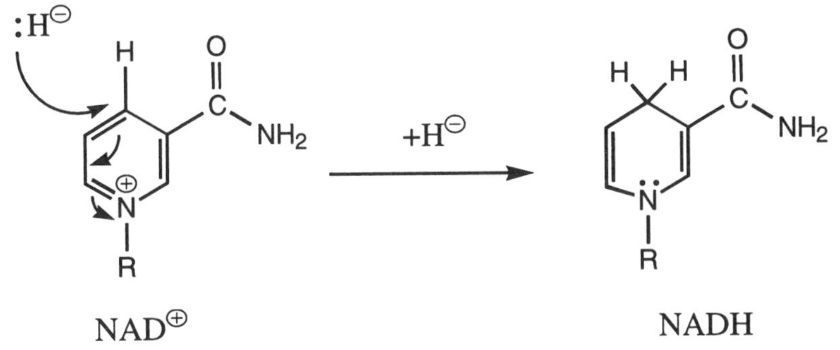

Figure 8. Reduction of $NAD^{\ominus}$ to NADH

Among biological oxidizing agents, $NAD^{\oplus}$ is relatively weak. For example, the equilibrium constant for an $NAD^{\oplus}$-dependent oxidation is about 10^{38} times smaller than that for the same oxidation by O_2, one of the cell's most powerful oxidizing agents. It takes a powerful reducing agent like the aldehyde of G3P to reduce $NAD^{\oplus}$ to NADH. Of course, the reduced form of a weak oxidizer is a strong reducer, and reduced nicotinamides like NADH act as powerful reducing agents in the synthesis of many biomolecules. The reducing power of NADH is due at least in part to the fact that, when NADH acts as a reducing agent, the product, $NAD^{\oplus}$, is **aromatic**, and hence particularly stable. Notice in Figure 8 that the nicotinamide ring of $NAD^{\oplus}$ contains six atoms and three alternating double bonds. You can therefore write two equivalent resonance contributors for this ring, suggesting that it is significantly stabilized by resonance, and suggesting further that the *p*-orbitals of the six sp^2-hybridized atoms of the ring form a continuous, cyclic π system containing six delocalized electrons. Quantum mechanical models suggest that special stability attends such cyclic systems when the number of delocalized electrons can be expressed as $4n + 2$, in which n is any whole number. Compounds with this electronic configuration (2, or 6, or 10, ... π electrons) are called **aromatic**. The most familiar aromatic compound is benzene (C_6H_6), but aromatics can also contain **heteroatoms** (non-carbon atoms), such as the nitrogen in the **heteroaromatic** ring of $NAD^{\oplus}$.

In Reaction 7 of Figure 1, the phosphate of 1,3-BPG is transferred to ADP to form ATP. The mechanism is similar to the reverse of Reactions 1 and 3, but note that 1,3-BPG is a mixed anhydride (carboxylate-phosphate) and not a phosphoester like G6P or FBP. Because mixed anhydrides are more reactive than phosphoanhydrides, transfer of phosphate from 1,3-BPG to ADP is favorable. The reaction is reversible, however, and serves as a step in the synthesis of glucose, as well as its degradation. The product of the forward reaction, ATP, is a source of chemical energy that is used to drive many cellular processes.

Reaction 8, the conversion of 3PG to 2PG, appears to be simply a transfer of the phosphate group from one hydroxyl (at C-3) to another (at C-2). This is an example of **transesterification**. Figure 9a shows a typical organic transesterification, in which the acyl group of an ester is transferred to an alcohol, leaving a different alcohol behind. See if you can fill in the mechanistic details of this reaction, using a hydrogen ion as the catalyst. The first step is protonation of the carbonyl oxygen.

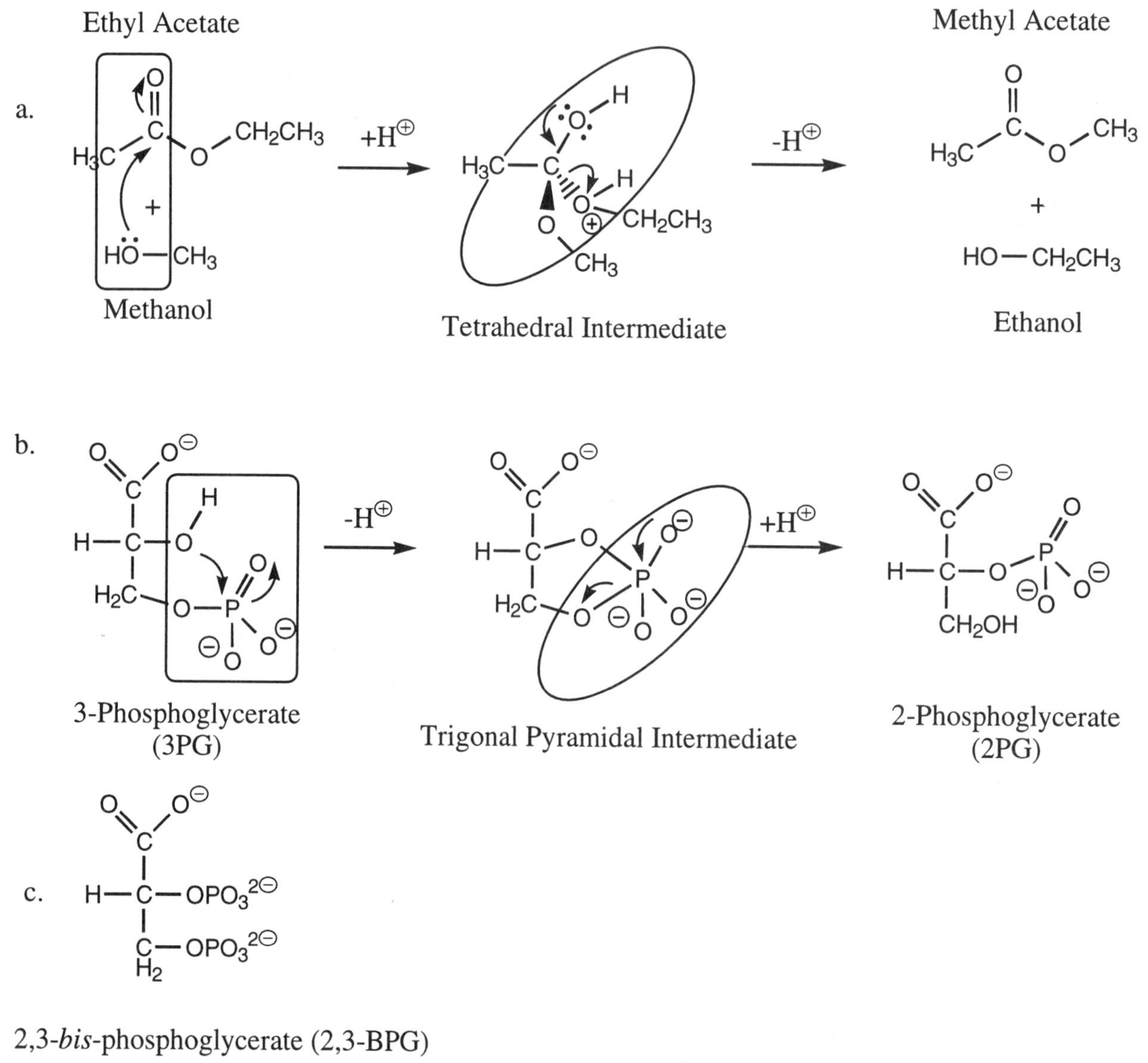

Figure 9. Transesterification.   **a.** Conversion of Ethyl Acetate to Methyl Acetate.
b. Conversion of 3PG to 2PG: A Plausible, But Incorrect Mechanism.   **c.** 2,3-DPG.

Figure 9b shows a plausible mechanism for Reaction 8. Analogous intermediates in Figures 9a and 9b are circled. When you study glycolysis, you will learn that this reaction is more complex than it appears here, and that the phosphate is not simply transferred directly from C-3 to C-2. As evidence of a more complex mechanism, in yeast and muscle, each molecule of the enzyme that catalyzes this reaction is known to contain one phosphate group. If the phosphate groups of the enzymes are labeled with the radioisotope ^{32}P, the radioactive phosphorus appears in the phosphate of 2PG, suggesting that the enzyme swaps the phosphate group of 3PG for its own phosphate group. In fact, a dephosphorylated, inactive enzyme can be restored to full activity by a stoichiometric amount of 2,3-*bis*-phosphoglycerate (2,3-BPG, shown in Figure 9c), suggesting that 2,3-BPG is an intermediate in the reaction. This is another example of the power of radioisotopic labeling in revealing details of reaction mechanisms.

Reaction 9 in Figure 1 is an example of **elimination**. The elements of one water molecule, a hydrogen from C-2 and the OH from C-3, are eliminated from 2PG to produce PEP. Figure 10 provides a mechanism, as if catalyzed by $H^{\oplus}$. Again notice the use of curved arrows in this mechanism. In step 1, the curved arrow shows protonation of O-3, with an unshared electron pair of the oxygen becoming a bonding pair in the $—OH_2^{\oplus}$ group. In step 2, the hydrogen atom of C-2 is removed by a base, and its bonding electron pair becomes a second bonding pair, making a double bond between C-2 and C-3. This new bond displaces a water molecule, which is a much better leaving group than was the C-3 OH before it was protonated. In this mechanism, notice that one proton is taken up (step 1) and one is released (step 2), so this process would not alter the concentration of hydrogen ions. Thus we can say that $H^{\oplus}$ is truly a **catalyst** in this process, because its concentration is not changed by the overall reaction.

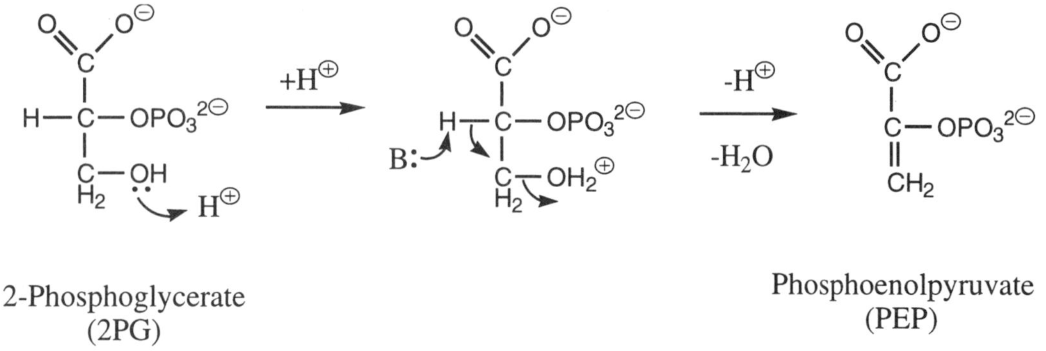

2-Phosphoglycerate
(2PG)

Phosphoenolpyruvate
(PEP)

Figure 10. Acid-Catalyzed Elimination of Water.

In Reaction 10 of Figure 1, another phosphate transfer occurs, from O-2 of PEP to ADP to form ATP and pyruvate. This process looks at first glance like conversion of a phosphate ester (PEP) to an anhydride, which you would not expect to be favorable according to the reactivity series. The reason that this reaction is favorable is that the initial product of the phosphate transfer is not pyruvate, but enolpyruvate, which quickly tautomerizes to pyruvate. In this keto-enol equilibrium, pyruvate, like most keto forms, is greatly favored. Remember that when one or more equilibria occur in sequence, the overall equilibrium constant is the product of the equilibrium constants for the individual steps. The large equilibrium constant for conversion of enolpyruvate to pyruvate makes the overall equilibrium favorable for Reaction 10, although the phosphate transfer itself may be much less favorable. Another way to picture this is to note that the direct product of the phosphate transfer, enolpyruvate, is removed immediately by tautomerism, which drives the reaction to completion. **Le Chatelier's principle**, which states that equilibria spontaneously shift in the direction that minimizes changes in concentrations of reactants or products, captures the logic of this view. Recall that many chemical reactions go to completion because products, for instance gases or precipitates, are lost from solution when they form.

In a living organism, what happens to the pyruvate that results from glycolysis? Like all the compounds in Figure 1, pyruvate is a metabolic intermediate. It occupies a branch point in metabolism, and can enter several other pathways. Some cells break down pyruvate further, ultimately to CO_2, in oxidative processes that require the participation of oxygen from the air. Others, like muscle cells during heavy exercise, reduce pyruvate to lactate ($CH_3CHOHCOO^{\ominus}$). The overall reaction

$$C_6H_6O_6 \rightarrow 2\,C_3H_5O_3{}^{\ominus} + 2\,H^{\oplus}$$

acidifies muscle and causes cramps. Still others cells, like yeasts, convert pyruvate to ethanol (CH_3CH_2OH) and CO_2. This two-step process begins with the **decarboxylation** of pyruvate to form acetaldehyde (CH_3CHO), which is then reduced to ethanol. The loss of CO_2 drives this process to completion. Both lactate and ethanol production entail **reduction** by NADH, and this reduction exactly balances the oxidation in Reaction 6. The conversion of glucose to lactate or to ethanol and CO_2 therefore entails no net oxidation or reduction. (To convince yourself of this, calculate the average oxidation number for carbon in glucose and in lactate.) Because such redox-balanced processes can be sustained in the absence of oxygen, they are referred to as *anaerobic*. Indeed, anaerobic (oxygen-free) conditions are necessary for production of ethanol by yeasts during manufacture of beer and wine. If air is present, yeasts oxidize glucose to CO_2 and produce no ethanol.

Pyruvate can also be converted to the **α-amino acid** alanine (Figure 11a), one of the building blocks required for the synthesis of proteins.

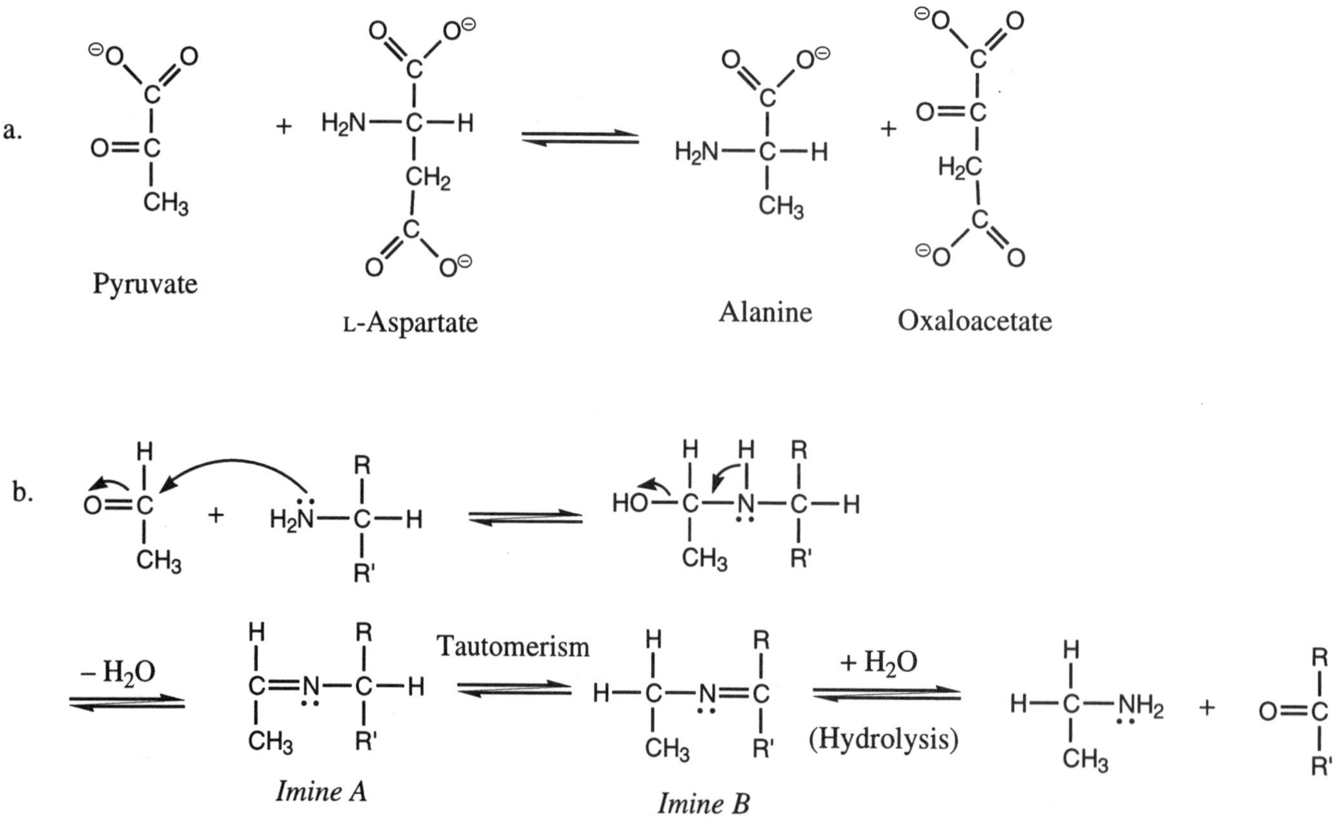

Figure 11. Transaminations **a.** Formation of Alanine from Pyruvate **b.** General Mechanism of Transamination

One of the paths from pyruvate to alanine provides another important example in this review, the reaction of amines with ketones (or aldehydes) to form **imines** (**Schiff bases**). Pyruvate, an α-ketoacid, receives its amine group in a **transamination** reaction with an α-amino acid, which in turn becomes an α-ketoacid. Figure 11a shows this process, with the amino acid aspartate as the source of the amino group, and oxaloacetate as the by-product. Figure 11b shows with general structures how a ketone and an amine can exchange an amine group. First, they join to form an imine (Figure 11b, imine A) in a **condensation** (water-forming) reaction. This process is reversible, and the reverse reaction, hydrolysis of imine A, would regenerate the reactants. But imine A equilibrates (another example of tautomerism) with imine B, which is also subject to hydrolysis. Hydrolysis of imine B completes the transfer of the amino group from one reactant to the other. Try writing a mechanism for the hydrolysis of imine B, by analogy with the reverse of the initial condensation that forms imine A. Biological transaminations are more complex than shown in Figure 11, involving a third amine carrier that receives the amine group from an α-amino acid and then transfers it to an α-ketoacid. Both transfers entail a mechanism similar to that shown in Figure 11b. Imines that are subject to tautomeric shifts that move the double bond from one side of the nitrogen to the other are instrumental in a number of group transfers like transamination.

As mentioned earlier, α-amino acids like alanine are the building blocks of proteins. Proteins are **polymers** constructed from α-amino acids that are joined into long chains by condensation reactions. The process is complex, but the overall reaction is the condensation of the α-amino group of one amino acid with the α-carboxyl group of the next, so that the two are joined by an **amide** linkage, which in proteins is called a *peptide bond*. When two amino acids are joined by a peptide bond (Figure 12a), the product is called a *dipeptide*. Water is a byproduct. There are twenty different α-amino acids that the cell can use in protein synthesis, and proteins range in size from less than 100 to more than 1000 amino-acid residues. Figure 12b shows a portion of a protein chain, with

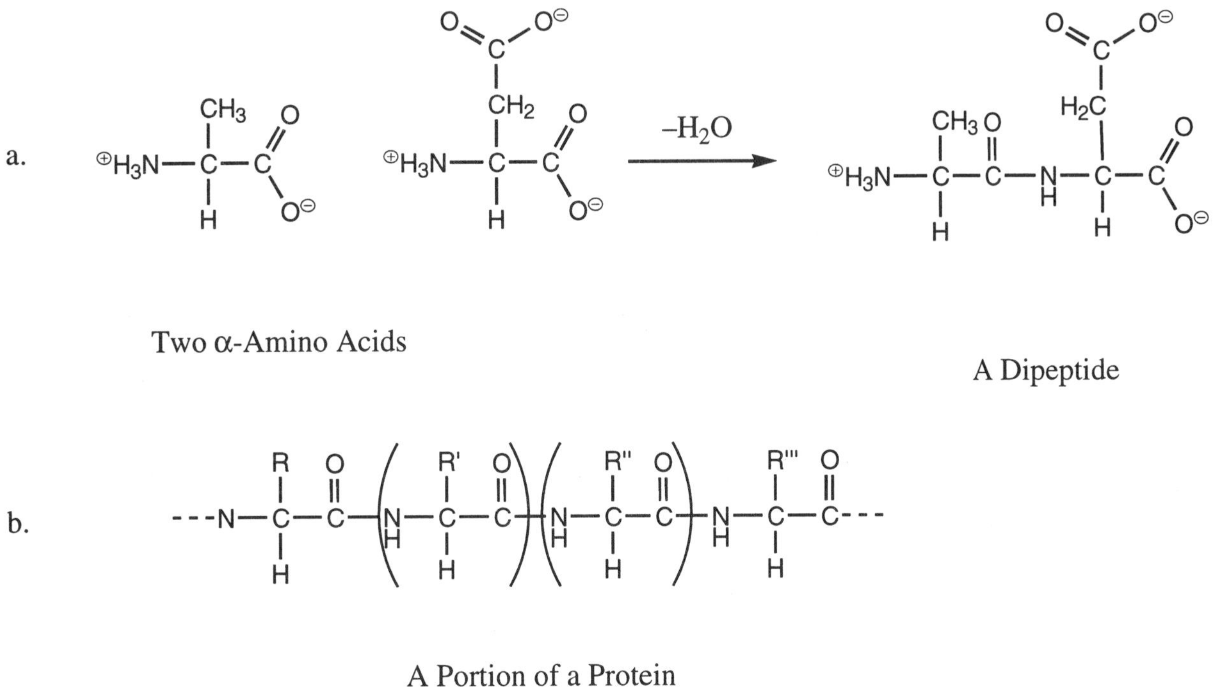

a.

Two α-Amino Acids

A Dipeptide

b.

A Portion of a Protein

Figure 12. Amides   **a.** Condensation of Two α-Amino Acids to Form a Dipeptide   **b.** Proteins are Polymers of α-Amino Acids

the *residues* (the remains after loss of water) of two amino acids marked by parentheses. The potential number of different proteins containing 100 amino acids is 20^{100} (or about 10^{130}), since each of the 100 positions in the chain can be any one of 20 different amino acids. The universe contains far fewer than 10^{130} atoms (about 10^{80}). It is no wonder that cells can make a bewildering array of proteins from a small number of building blocks. Each enzyme in glycolysis, as well as each enzyme in every other metabolic pathway, is a protein. Even relatively simple bacterial cells contain thousands of different proteins.

Enzymes Versus Chemical Catalysts

Let me remind you that all the reactions in glycolysis are catalyzed by enzymes (each of which is a protein), and that the mechanisms of the reactions are more complex than I have shown here. In these examples, I have simply written *plausible* mechanisms to provide a review of basic skills like electron pushing. Remember that a chemist proposes a mechanism to explain all available experimental evidence concerning a reaction. Realistic mechanisms, like the ones you find in your biochemistry text, are acceptable only if they satisfy a number of requirements. For instance, they must specify a role for all reagents known to be essential to the reaction, including specific parts of the enzyme; they must include all intermediates that have been detected in the reaction, for example, by spectroscopy or chemical trapping; and they must be compatible with results of isotopic labeling experiments, with the stereochemical consequences of the reaction, and with measurements of reaction **kinetics** (dependence of reaction rates on reaction conditions). In short, a mechanism is an attempt to explain the course of a reaction in a way that accounts for all that is known about the process. The quest to determine the mechanisms of biological reactions lifts mechanistic organic chemistry to its greatest heights.

Enzymes are far more powerful and selective agents than chemical agents like hydrogen ions. As an example, consider Reaction 1 again. Any of the five hydroxyl groups of D-glucose could just as readily accept the phosphate group from ATP; all five hydroxyls have about the same chemical properties. But the enzyme that catalyzes this reaction <u>directs</u> the new phosphate group to C-6 only, despite the almost identical reactivity of the other hydroxyls. This is an example of **regiospecificity** in enzymatic catalysis: enzymes can distinguish chemically identical but structurally distinct functional groups, and thus enzymes do not produce the side products that plague laboratory reactions.

As a further example, notice that Reaction 4, like most of the reactions in glycolysis, is a reversible reaction. This means that in the reverse reaction, the enzyme assembles FBP from DHAP and G3P. The process in this direction produces two new chiral centers: C-3 and C-4 of FBP. In this enzyme-catalyzed aldol condensation, only one of four possible diastereoisomers is produced: FBP itself. Any <u>chemically</u> catalyzed version of this reaction would produce all four products (the chiral center of G3P would result in unequal amounts of the four diastereomers, an example of **asymmetric induction**). The enzyme that catalyzes Reaction 4 splits only FBP in the forward direction, and produces only FBP in the reverse direction. This is an example of **stereospecificity** in enzymatic catalysis: enzymes can distinguish stereoisomers, even enantiomers, as reactants, and can produce specific stereoisomers as products, again without the side products that are inevitable in most laboratory reactions. No wonder that biochemists are seeking ways to use enzymes as commercial catalysts.

An understanding, at the molecular level, of the power of enzymes is just one of many interesting insights that await you in your study of biochemistry. You are embarking on a study of one of the most powerful, fast-moving, and exciting areas of science. I hope that this review of important concepts from organic chemistry helps you to get off on the right foot in your biochemistry course.

2

Water

_____

I. KEY TERMS

With the help of this study guide, your textbook, and class notes, you should be able to define and explain the significance of the following terms:

polar	osmotic pressure	nucleophile
hydrogen bond	hydrophobic	electrophile
electrolyte	hydrophobic effect	acid
hydrophilic	amphipathic	base
cation	micelle	pH
anion	chaotrope	conjugate base
solvation sphere	charge-charge interaction	conjugate acid
solvated	ion pairing	pK_a
hydrated	van der Waals forces	Henderson–Hasselbalch equation
osmosis	hydrophobic interaction	buffered

II. EXERCISES

A. True–False

_____ **1.** The O–H bonds in water are polar due to the high electronegativity of hydrogen.

_____ **2.** Water is a polar molecule (has a dipole moment) because it is V–shaped.

_____ **3.** In the liquid state, each water molecule has the potential to form hydrogen bonds with four other water molecules.

_____ **4.** Compounds that dissolve readily in water are termed hydrophilic and include electrolytes as well as nonelectrolytes.

_____ 5. The osmotic pressure of an aqueous 0.001 M starch solution is greater than that of an aqueous 0.001 M glucose solution.

_____ 6. Polar molecules can induce polarity in nonpolar molecules.

_____ 7. An aqueous solution of pH 3.0 contains H_2O, $H^\oplus$, and $OH^\ominus$.

_____ 8. An acid that dissociates to the extent of 88% in water would be termed a strong acid.

_____ 9. A solution that contains 0.05 mol of lactic acid and 0.05 mol of potassium lactate per liter of solution has a pH of 3.86.

_____ 10. The major buffer system of the blood is the bicarbonate–carbonic acid buffer system.

_____ 11. The shell of water molecules that surrounds a dissolved ion or molecule is called a hydrosphere.

B. Short Answers

1. _____ is the tendency of an atom to attract to itself the shared electrons in a covalent bond.

2. _____ agents are certain ions and molecules that are poorly solvated in water and enhance the solubility of nonpolar compounds by disordering the water molecules.

3. _____ molecules contain both hydrophobic and hydrophilic groups.

4. Soap molecules dissolved in water tend to form aggregates called _____.

5. A shell of water molecules that forms around ions as they become dissolved is called a/an _____.

6. Attractive forces between molecules, collectively called _____, can occur due to dipole-dipole attractions, ion-dipole attractions, and interactions between two non-polar atoms or molecules which result in transiently induced dipoles and which are also known as London dispersion forces.

7. The unit(s) of K_w, the ion-product constant of water, is(are) _____.

8. An acid or base that dissociates less than 100% in water is described as being _____.

9. The unit(s) of K_a, the dissociation constant for a weak acid, is(are) _____.

10. A solution that contains equal or nearly equal quantities of a weak acid and its conjugate base is called a/an _____.

11. The condition called _____ results when a person's blood pH becomes lower than normal.

12. Compounds that dissolve in water to produce ions that can conduct a current through the resulting solution are called _____.

C. Problems

1. Calculate the pH of (a) a solution of 0.0100 M acetic acid and (b) a solution of 0.0100 M HCl. How do they compare?

2. In a solution of 0.00250 M HNO_3, evaluate $[H^{\oplus}]$, pH, and $[OH^{\ominus}]$.

3. In an acetic acid–acetate buffer of pH 4.00, what is the ratio of acetate to acetic acid?

4. Your text indicates that the energy required to break each hydrogen bond in ice is 23 kJ/mol (section 2.2) and that four hydrogen bonds are possible for each H_2O molecule. Assuming this bond energy is the same for ice and liquid water at 0.0°C, but that 78.2 kJ/mol is the actual energy required to break the hydrogen bonds in the liquid water, determine the average number of hydrogen bonds per molecule that exist in the liquid water in contrast to the four per molecule in ice. Express your answer as a number such as 1.7 or 2.6 or 3.9, etc. bonds per water molecule.

5. Write the formula of the species which, together with H_2O, constitutes a conjugate acid–base pair. What does your answer say about the properties of H_2O?

6. Here is a thought question some instructors have used: What is the pH of a 10^{-8} M HCl solution?

7. What is the pH of 0.0112 M lactic acid?

8. During the titration of 50.0 ml of 0.0500 M lactic acid with 0.0500 M NaOH, the biochemist stopped after 25.0 ml of the base had been added. Can this solution be used as a buffer? What is the pH of the solution?

9. What is the pH of a solution containing 0.0450 M benzoic acid and 0.0400 M sodium benzoate? The pK_a for benzoic acid is 3.19.

10. At the normal physiological pH of 7.4, some sources indicate that the ratio of bicarbonate to carbonic acid in the blood is about 20:1. This would not appear to be an ideal buffer since the two buffer components are not in approximately equal quantities. How is this system able to adequately neutralize $OH^{\ominus}$?

D. Additional Problems

1. Refer to Problem C.1. in which you contrasted the pH of 0.0100 M HCl to that of 0.0100 M acetic acid. From that information and the solution to that problem, calculate the % ionization of acetic acid in that solution.

2. If the ratio of bicarbonate to carbonic acid is 20:1 in normal blood, what does this information indicate is the pK_a for carbonic acid? How does this compare to the value in text Table 2.4?

3. At 25°C, water ionizes such that the value of K_w is 1.00×10^{-14}. At higher temperatures, a greater degree of ionization will occur. If $K_w = 1.4 \times 10^{-14}$ at some temperature greater than 25°C, what is the pH at this elevated temperature? Is the system acidic, basic, or neutral?

4. What is the pH of a solution prepared by mixing 600.0 ml of 0.075 M KH_2PO_4 with 400.0 ml of 0.12 M K_2HPO_4?

5. If the pH of distilled water in equilibrium with air is about 5.68, what is the maximum possible concentration of dissolved carbon dioxide? Use information in text Table 2.4.

3

Amino Acids and the Primary Structures of Proteins

I. KEY TERMS

With the help of this study guide, your textbook, and class notes, you should be able to define and explain the significance of the following terms:

zwitterion

stereoisomer

disulfide bridge

hydropathy

cathode

microenvironment

peptide bond

C-terminus

column chromatography

HPLC

gel-filtration chromatography

ligand

amino acid analysis

Edman degradation procedure

chiral

configuration

enantiomer

isoelectric point

anode

primary structure

N-terminus

fractionation

eluate

ion-exchange chromatography

affinity chromatography

electrophoresis

polyacrylamide gel electrophoresis

sequenator

II. EXERCISES

A. True–False

_____ **1.** In a medium of pH 2.0, aspartic acid has a net positive charge.

_____ **2.** There is no pH condition at which the predominant form of lysine would be a species with three separate charges.

_____ **3.** The majority of the amino acids used to make natural proteins have highly hydrophilic R-groups.

_____ 4. One mole of aspartic acid in its fully protonated form would require three moles of OH$^\ominus$ to convert it to the fully unprotonated form.

_____ 5. All 20 of the amino acids used to make natural proteins are optically active.

_____ 6. The isoelectric point for most of the amino acids having non-polar side chains is about 6.

_____ 7. Since the common amino acids have the L- stereo configuration, one may assume that they will rotate plane polarized light in a left-handed (levorotatory) direction.

_____ 8. The addition of 0.015 mole of L-valine to a solution containing 0.015 mole of D-valine would produce a solution that would exhibit twice the degree of rotation of polarized light that the D-valine solution had before the addition.

_____ 9. A linear molecule made from amino acids linked end to end, which has four peptide bonds in its structure, is called a tetrapeptide.

_____ 10. All of the α-L-amino acids with only one chiral carbon have the (S) designation in the RS system.

_____ 11. When electrophoresis is performed on a mixture of methionine and aspartic acid, buffered at pH 5.7, methionine will move toward the cathode and aspartic acid will move toward the anode.

_____ 12. In SDS-PAGE, proteins are separated from each other primarily on the basis of their size and not their charge.

B. Short Answers

1. In partition column chromatography, one or more solutes may be retained on the column while another passes through with little or no retention. When a solute is removed from the column, it is said to have been _____.

2. An amino acid in a dipolar ion form can be called a/an _____.

3. The pH at which a given amino acid carries a net zero charge is referred to as its _____.

4. The covalent linkage formed by the oxidation of the R-groups of two cysteine residues in a peptide or protein is called a/an _____.

5. A measure of the relative hydrophobicity or hydrophilicity of a given amino acid is called its _____.

6. The immediate surroundings of a side chain of a given amino acid residue in a protein is called its _____.

7. D-leucine and L-leucine may be called a pair of _____.

8. The specific sequence of amino acid residues in a peptide or protein is called its _____.

9. The _____ can be used to determine the sequence of a peptide by means of an automated instrument called a/an _____.

10. The sequence of a protein or peptide is given direction by normally indicating its primary structure from its _____ to its _____.

C. Problems

1. Of 100 molecules of L-alanine, how many have ionized carboxyl groups at pH 3.0?

2. Stereoisomers that are not enantiomers are called diastereomers. Draw a diastereomer of the form of α-L-threonine shown here.

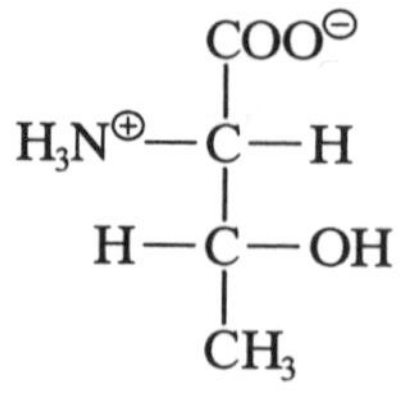

$$
\begin{array}{c}
COO^{\ominus} \\
| \\
H_3N^{\oplus}-C-H \\
| \\
H-C-OH \\
| \\
CH_3
\end{array}
$$

3. Why is glycine a crystalline solid rather than being a liquid as are many other organic molecules of similar molecular weight?

4. Draw the complete structure of glycylvalylalanine at pH 7.

5. Draw the ionic forms of L-serine that predominate at each pH:
(a) 2.0 (b) 5.7 (c) 10.0

6. Explain the electrophoretic migration behavior you would expect of aspartame at (a) pH 3.0 and (b) pH 6.9. Assume that the pK_a values of the free amino acids given in text Table 3.2 are applicable for this dipeptide.

7. Draw an enantiomer and a diastereomer of this molecule.

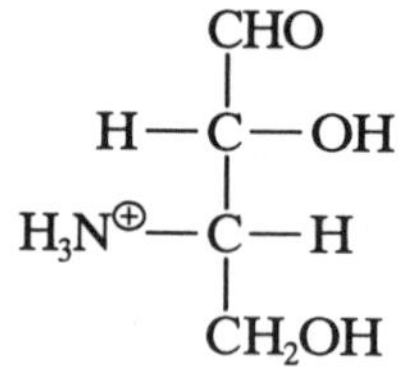

$$
\begin{array}{c}
CHO \\
| \\
H-C-OH \\
| \\
H_3N^{\oplus}-C-H \\
| \\
CH_2OH
\end{array}
$$

8. Calculate the isoelectric points of
(a) glycine (b) L-lysine (c) L-aspartic acid

9. Use graph paper to draw the titration curve (pH on the y-axis vs. equivalents of $OH^{\ominus}$ on the x-axis) for one mole of an amino acid with $pK_{a1} = 2.4$ and $pK_{a2} = 9.6$. Label these points on the curve:
(a) pK_{a1} (α-carboxyl group)
(b) pK_{a2} (α-amino group)
(c) pI (isoelectric point)

10. Assign the appropriate *RS* designation to this molecule of 2-amino-4-bromo-4-hydroxybutanoic acid.

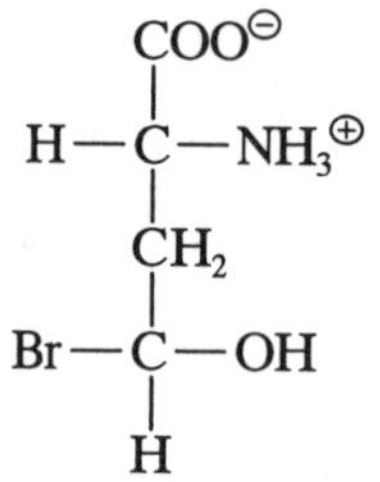

$$
\begin{array}{c}
COO^{\ominus} \\
| \\
H-C-NH_3^{\oplus} \\
| \\
CH_2 \\
| \\
Br-C-OH \\
| \\
H
\end{array}
$$

D. Additional Problems

1. Calculate the ratio of conjugate base to acid form for each of the ionizable groups of L-glutamic acid in a medium of pH 4.00.

2. How many different peptides of 15 residues can be made from the 20 common amino acids?

3. You have a mixture of L-lysine, L-aspartate, and glycine that is subjected to electrophoresis in a buffered solution of pH 6. How will the amino acids behave? Explain.

4. The labels have been lost from two dipeptide samples. One is known to contain a residue each of L-alanine and glycine. The other contains a residue each of L-alanine and L-aspartic acid. A 1.0 mmole sample of one of the dipeptides in its fully protonated form required 30. ml of 0.10 M NaOH to convert it to its totally unprotonated form. Which dipeptide is it? Explain.

5. A small peptide was found to contain equimolar amounts of the following amino acids: L-arginine, L-glutamic acid, glycine, L-lysine, L-methionine, and L-phenylalanine. Individual samples of the peptide were treated with the following agents with the results noted:

    (a) trypsin: L-arginine and a pentapeptide.
    (b) cyanogen bromide: two tripeptides.
    (c) *S. aureus* V8: L-lysine and a pentapeptide.
    (d) chymotrypsin: a dipeptide and a tetrapeptide, with the latter showing absorbance at 260 nm.

    What is the primary structure of the peptide? Explain each piece of evidence given and the reasoning that led to your answer.

4

Proteins: Three-Dimensional Structure and Function

I. KEY TERMS

With the help of this study guide, your textbook, and class notes, you should be able to define and explain the significance of the following terms:

conformation	positive cooperativity of binding	β-strand
globular protein	allosteric interaction	loop
primary structure	allosteric protein	hairpin loop
tertiary structure	T state	domain
peptide group	Bohr effect	oligomer
pitch	antibody	cooperativity of folding
capping box	immunoglobulin fold	denaturation
β-structure	immunoassay	Schiff base
β-sheet	native conformation	oxygenation
turn	fibrous protein	allosteric modulator
motif	secondary structure	R state
monomer	quaternary structure	2,3-*bis*phosphoglycerate (BPG)
multienzyme complex	α-helix	carbamate adduct
molecular chaperone	rise	antigen
random coil	3_{10} helix	hypervariable region

II. EXERCISES

A. True–False

_____ **1.** Secondary structural features of a protein are stabilized by covalent bonds.

_____ **2.** Nearly all peptide groups in proteins studied to date have been found to have the *trans* conformation.

_____ **3.** The tertiary structure of a globular protein is its overall three-dimensional shape.

_____ **4.** Oxygen-binding characteristics of hemoglobin proved to be quite different from those of myoglobin due to radical differences in their respective protein chains.

_____ **5.** The rise per amino acid residue in a segment of α-helix is called the pitch.

_____ **6.** The heme group of hemoglobin binds oxygen more strongly alone than it does when present in the hemoglobin molecule.

_____ **7.** Recognizable combinations of α-helices and β-strands that appear in a number of different proteins are called domains.

_____ **8.** The hydrophilic side chains of amino acid residues normally locate themselves toward the exterior of globular proteins.

_____ **9.** Denaturation destroys the biological activity of a protein by breaking many covalent bonds in the protein.

_____ **10.** The quaternary structure of proteins is maintained predominately by non-covalent interactions.

_____ **11.** Hemoglobin binds oxygen more strongly as the pH is lowered.

B. Short Answers

1. The decreased affinity of hemoglobin for oxygen due to elevated levels of carbon dioxide and $H^{\oplus}$ is called _____.

2. The type of secondary structure involving 75% of the residues in myoglobin is _____.

3. Two modes of denaturation that would primarily affect the salt bridges and disrupt hydrogen bonding, respectively, in a protein are _____ and _____.

4. The form of the protein chain that occurs upon denaturation is called _____.

5. Proteins that possess quaternary structure are called _____.

6. The actual site of oxygen binding in the heme of myoglobin is _____.

7. A/an _____ is the term for an oligomeric protein responsible for catalyzing several different metabolic reactions.

8. The process by which a protein in a random coil conformation assumes its native shape is called _____.

9. The sigmoidal curve for the binding of oxygen by hemoglobin illustrates the phenomenon of _____.

10. An organic molecule in erythrocytes that lowers the affinity of hemoglobin for oxygen is _____.

11. _____ are proteins that assist in the cellular process of protein folding such that a large fraction of the protein molecules affected achieve the native conformation and are thus biologically active.

12. A region in a globular protein characterized by negative Φ and Ψ values is a region of _____.

13. In the immune response, foreign compounds called _____ cause the synthesis of proteins called _____, which combine with and precipitate the foreign compounds and mark them for destruction.

C. Problems

1. A segment of a protein chain in the α-helical conformation contains 20 residues. How long is this segment?

2. How many turns are in this helical segment of 20 amino acids?

3. Explain the manner in which hydrogen bonding occurs in a region of α-helix.

4. Consider the hydrogen bonding in an α-helical segment of 10 residues. Considering the potential for hydrogen bond formation, what percentage of these potential hydrogen bonds actually exist? For the purpose of this question, assume that all atoms of all residues are "in" the helix (actually, there is evidence that not all atoms of the residues at the ends of a helical segment are in the helix).

5. Consider this segment of protein chain and speculate on the length(s) of α-helix that is/are feasible.

−Asp–Met–Phe–Pro–Ala–Met–Ala–His–Leu–Ile–Pro–

6. Scurvy is a human disease characterized by skin lesions, loose teeth, and bruises on arms and legs due to fragile blood vessels. What is the connection between scurvy, vitamin C, and collagen?

7. The enzyme RNase A is denatured, and rendered inactive, by treatment with 8 M urea containing 2-mercaptoethanol. If both urea and the 2-mercaptoethanol are removed by dialysis, and the denatured protein is exposed to air, biological activity is regained. Alternatively, if the denatured protein is reoxidized in the presence of urea, only about 1% of the original activity is regained. What conclusion(s) might be made from these observations. Explain.

8. Relatively small changes in pH have rather large effects on the binding of oxygen by hemoglobin. If the pO_2 is 30 torr in certain tissues, is more, or less, oxygen bound at pH 7.6 compared to that bound at pH 7.2? Express the amount of oxygen bound at pH 7.6 versus that at pH 7.2 as a ratio.

9. Your text indicates that some carbon dioxide is carried by hemoglobin by the formation of carbamate adducts with $-NH_2$ groups of hemoglobin. Most CO_2 is transported as dissolved bicarbonate ions.

(a) Write the reaction for the formation of carbamate adducts.

(b) Indicate how dissolved bicarbonate ions are involved in the eventual release of CO_2 at the lungs.

(c) Speculate on the importance of the carbamate adduct formation to transport CO_2 in terms of the overall amount of CO_2 that must be transported.

10. What is an immunoassay? Explain the underlying basis for this technique that allows it to be used, in some cases, for diagnostic tests.

D. Additional Problems

1. The energy required to break a carbon-carbon bond is about 345 kJ/mol (Holtzclaw, General Chemistry, 9th Edition, 1991, page 194). Hydrogen bonds in proteins help stabilize secondary and tertiary structure. How many hydrogen bonds are required to provide the stabilization that would be the energy equivalent of one carbon-carbon bond?

2. The enzyme trypsin contains six disulfide bonds. In one study, renaturation of the denatured enzyme, following removal of the denaturant, yielded only 8% of its original activity. (a) What would be the expected random reformation of the six disulfide bridges? (b) Why is essentially full activity not regained as it was with RNase A?

3. Calculate the stabilization energy due to hydrogen bonding in the myoglobin molecule.

4. What is the percent of extensibility of keratin in hair?

5. What is the percent of deoxymyoglobin (Mb) in a solution where the partial pressure of oxygen, pO_2, is 6.0 torr? The P_{50} for myoglobin is about 2.8 torr.

5

Properties of Enzymes

I. KEY TERMS

With the help of this study guide, your textbook, and class notes, you should be able to define and explain the significance of the following terms:

enzyme
substrate
active site
dehydrogenase
transferase
lyase
ligase
synthetase
rate equation
kinetic order
second order reaction
enzyme-substrate complex
saturation
initial velocity
Michaelis-Menten equation
specificity constant, k_{cat}/K_m
Lineweaver-Burk plot
kinetic mechanism
ordered mechanism

ping pong reaction
inhibition constant
reversible inhibitor
uncompetitive inhibition
affinity label
regulatory enzyme
allosteric modulator
sequential theory
converter enzyme
enzymatic reaction
reaction specificity
product
oxidoreductase
hydrolase
isomerase
synthase
velocity
rate constant
first order reaction

zero order reaction
enzyme assay
catalytic constant
maximum velocity
Michaelis constant, K_m
rate acceleration
turnover number
sequential reaction
random mechanism
inhibitor
irreversible inhibitor
competitive inhibition
noncompetitive inhibition
site-directed mutagenesis
regulatory site
concerted theory
interconvertible enzyme
covalent modification

II. EXERCISES

A. True–False

_____ **1.** The only biological catalysts are the proteins called enzymes.

_____ **2.** All enzymes require cofactors such as FAD and coenzyme A.

_____ **3.** An enzyme that catalyzes the addition of a phosphoryl group to glucose would probably not catalyze the same reaction for glycerol.

_____ **4.** Any form of contact between an enzyme and its substrate will lead to reaction.

_____ **5.** The formation of an enzyme-substrate complex is an unproven hypothesis.

_____ **6.** A multisubstrate, enzyme-catalyzed reaction in which a product is released before all substrates have been bound to the enzyme is an example of a sequential kinetic mechanism.

_____ **7.** K_m and V_{max} values can best be determined graphically from a plot of initial velocity, v_0, versus substrate concentration, [S]. This type of plot is known as a Michaelis-Menten plot.

_____ **8.** The higher the K_m value, the greater the affinity of an enzyme for its substrate.

_____ **9.** The K_m values of enzymes for their substrates are usually slightly higher than the intracellular concentrations of their substrates.

_____ **10.** A plot of kinetic data, v_0 vs. [S], that produces a sigmoidal curve indicates that cooperative interactions occur between enzyme subunits.

_____ **11.** Regulatory enzymes are oligomeric proteins that catalyze reactions at a committed step in a pathway.

B. Short Answers

1. The region of an enzyme molecule with which the substrate must interact is called the _____.

2. The short-lived species formed when enzyme and substrate interact initially is the _____.

3. Enzymes that catalyze the conversion of a molecule into its structural isomer would belong to the IUBMB category of _____.

4. The number of catalytic events catalyzed per second per enzyme molecule (or per active site) is called the _____.

5. The ratio k_{cat}/K_m is called the _____ and is a measure of the _____.

6. Some enzymes are subject to control of activity by the addition and removal of phosphate groups. This is called regulation by _____.

7. Due to its structure, a/an _____ inhibitor binds to the active site of an enzyme.

8. An inhibitor that does not alter the K_m of the enzyme is a/an _____ inhibitor.

9. An inhibitor that alters both the K_m and V_{max} of the enzyme system is a/an _____ inhibitor.

10. Inhibition of a regulatory enzyme by an end product of the pathway is called _____ inhibition.

11. Sites where allosteric modulators bind to enzymes are called _____.

C. Problems

1. Sketch a graph of rate (velocity) vs. [E] for an enzyme-catalyzed system that contains saturating levels of substrate. Explain other necessary conditions for such a determination.

2. The enzyme alcohol dehydrogenase catalyzes the interconversion of ethanol and acetaldehyde and the interconversion of $NAD^{\oplus}$ and NADH. Start with acetaldehyde, the enzyme, and the appropriate form of the coenzyme and write the equation for this reaction, using the structures of the substrate, product, and the nicotinamide ring of the coenzyme.

3. If you wanted to follow the course of the reaction in the previous question by spectrophotometric techniques, how could it be done? Why would this be more difficult than following the reverse reaction in the same manner?

4. Your text indicates that K_m is the substrate concentration that gives a rate of $\frac{1}{2} V_{max}$. Show that a substrate concentration of 2 K_m does not give a rate of V_{max}. Do substrate concentrations of either 3 K_m or 4 K_m give a rate equal to V_{max}?

5. Why is the V_{max} of an enzyme-catalyzed reaction the same in the presence as it was in the absence of a competitive inhibitor?

6. How can one determine initial rates of reaction required for making Michaelis-Menten plots?

7. The K_m for the substrate of a particular enzyme is 2.0×10^{-5} M. If the initial velocity, v_0, is 0.16 μmol/min for [S] = 0.15, what will be the initial velocity when [S] = 2.0×10^{-4}?

8. Is a synthase the same as a synthetase? Explain.

9. Speculate on the advantage(s) for the cell of maintaining intracellular substrate concentrations within the range 0.1 K_m to K_m.

10. The following kinetic data were collected for an enzyme-catalyzed reaction. Determine (a) the V_{max} and (b) the K_m for this system.

Measurement Number	[S], M	v_0, μmol/min
1	1.0×10^{-5}	15.6
2	5.0×10^{-5}	34.6
3	1.0×10^{-4}	41.0
4	5.0×10^{-4}	47.9
5	6.0×10^{-2}	49.8
6	5.0×10^{-1}	50.0
7	8.0×10^{-1}	50.0

When a competitive inhibitor ([I] = 1.4×10^{-4}) was added to this system with [S] = 1.0×10^{-4}, a 30.0% decrease in initial velocity was observed. Find (c) the K_i for this competitive inhibitor. Note: It has been determined that $K_m^{app} = K_m (1 + [I]/K_i)$.

D. Additional Problems

1. In an enzyme-catalyzed reaction of the type:

$$2 A \longrightarrow B + C,$$

substrate concentrations of 5.0×10^{-1}, 5.0×10^{-2}, or 25.0×10^{-2} mM all produced 30. millimoles of product C in a 12-minute period. Express the level of enzyme assayed by this system in terms of International Units. Note: An International Unit of enzyme activity is defined as the number of micromoles of substrate transformed per minute by the enzyme.

2. If 2.0 mL was the volume of the enzyme solution used in the previous problem, and it contained 2.5×10^{-4} mg of total protein, what is the specific activity of the solution? Note: The specific activity of an enzyme preparation can be defined as the International Units of activity per mg of protein.

3. The enzyme in the previous two problems has a molecular weight of 12,500 g/mole and is thought to have just one active site per molecule. Based on the information in the previous two problems, what is the catalytic constant (turnover number) for this enzyme?

4. The kinetic data from an enzyme-catalyzed reaction is shown below. From this information, determine K_m and V_{max}.

Measurement Number	[S], M	v_0, μmol/min
1	1.7×10^{-6}	10.0
2	3.8×10^{-6}	20.0
3	1.2×10^{-5}	45.0
4	2.3×10^{-5}	60.0
5	8.5×10^{-5}	85.0

5. The enzymatic reaction of problem D. 4. is studied again in the presence of an inhibitor at 3.5×10^{-5} M concentration, giving the data shown below. Determine the type of inhibition and the K_i for the enzyme-inhibitor complex.

Measurement Number	[S], M	v_0, μmol/min
1	3.8×10^{-6}	12.2
2	1.2×10^{-5}	26.9
3	2.3×10^{-5}	36.8
4	8.5×10^{-5}	51.3

6

Mechanisms of Enzymes

I. KEY TERMS

With the help of this study guide, your textbook, and class notes, you should be able to define and explain the significance of the following terms:

mechanism

nucleophilic

carbanion

transition state

rate determining step

acid-base catalysis

diffusion controlled reaction

transition-state stabilization

thermodynamic pit

zymogen

catalytic triad

leaving group

electrophilic

carbonium ion

activation energy

effective molarity

catalytic center

covalent catalysis

proximity effect

catalytic antibody

trypsin inhibitor

low-barrier hydrogen bond

II. EXERCISES

A. True–False

_____ 1. A reactant species that is electron rich and attacks another reactant deficient in electrons is termed an electrophile.

_____ 2. When a carbon-carbon bond is split such that one carbon atom loses both electrons, that carbon atom becomes a carbonium ion.

_____ 3. A transition state proposed to form as part of a reaction mechanism is a transient species which cannot be trapped and studied.

_____ 4. The interior of globular proteins is generally a hydrophobic region. The active site of an enzyme is usually a cleft or pit in the globular protein. All the amino acid residues in the active site, therefore, have hydrophobic side chains.

_____ **5.** Ping-pong kinetics is a hallmark of enzymes that catalyze reactions involving covalent catalysis.

_____ **6.** Relatively strong binding of reactants in the enzyme active site is necessary for efficient catalysis.

_____ **7.** Entropy of reactants increases as reactants are bound by enzymes.

_____ **8.** An enzyme active site binds more tightly to a transition-state analog than to its substrate.

_____ **9.** Chymotrypsin is a proteolytic enzyme with three α-amino termini (N-termini).

_____ **10.** Serine proteases contain the same amino acid residues in the catalytic triad that is part of their active sites.

B. Short Answers

1. The _____ is a description of the bond-breaking and bond-forming events in a chemical reaction.

2. The _____ is the term for an energized arrangement of atoms in which bonds are being formed and broken prior to product formation in an enzyme-catalyzed reaction.

3. The energy required for reactants to reach the transition state from their ground state is called the _____ of the reaction.

4. In a _____ reaction, every collision between reactant molecules gives rise to product.

5. Rate enhancement by the binding of reactants close to each other in the enzyme active site is called the _____ effect.

6. Excessive stability of the enzyme-substrate complex, that is with substrate(s) tightly bound to the enzyme, is termed a/an _____.

7. Very strong hydrogen bonds that form when the electronegative atoms are less than 0.25 nm apart are termed _____.

8. In the catalytic triad of serine proteases, a _____ residue acts as a covalent catalyst.

9. _____ is the term applied to the catalytic mode that involves increased binding of transition states to enzymes as compared to binding of substrates or products.

10. The four major modes of enzymatic catalysis are: _____, _____, _____, and _____.

C. Problems

1. In the reaction of a certain anionic species, $Y:^{\ominus}$, with a reactant of the type R–C=O, would you expect

$$\overset{|}{X}$$

   (a) $Y:^{\ominus}$ to attack the carbon or the oxygen atom? Explain.
   (b) the leaving group to be a cationic or an anionic species? Explain.

2. What is the difference between intermediates and transition states?

3. Show the primary structure, using appropriate abbreviations for the amino acids, of the products formed by action of the following enzymes and chemical agent on this nonapeptide.

   Ala–Cys–Lys–Met–Phe–Arg–Ala–Tyr–Gly

   (a) elastase
   (b) trypsin
   (c) chymotrypsin
   (d) BrCN

4. Describe in a sentence or two how an enzyme causes a specific reaction to proceed more rapidly than the same un-catalyzed reaction.

5. If the serine proteases discussed in this chapter (trypsin, chymotrypsin, and elastase) contain the same catalytic triad, why do they individually catalyze cleavage of only those peptide bonds in which the carbonyl group is donated by specific amino acid side chains?

6. What is the function of enteropeptidase?

7. Suggest why trypsin can activate trypsinogen molecules to create more active trypsin.

8. Pancreatic trypsin inhibitor binds to and inactivates trypsin formed in the pancreas. Since trypsin inhibitor is itself a protein, why is it not digested by trypsin?

9. A certain proteolytic enzyme was inactivated with diisopropylfluorophosphate (DIFP). The inactive enzyme was subjected to enzymatic hydrolysis, which produced various peptides and the compound below. What does this evidence indicate about the enzyme?

$$(CH_3)_2CH-O-\overset{\overset{\displaystyle O}{\|}}{\underset{\underset{\displaystyle O-CH(CH_3)_2}{|}}{P}}-O-CH_2-\overset{\overset{\displaystyle NH_3^{\oplus}}{|}}{CH}-COO^{\ominus}$$

10. Summarize the way the catalytic triad of a serine protease achieves hydrolysis of a peptide bond. Consult Figure 6.27 in your text to help formulate your answer to this question.

D. Additional Problems

1. The maximum rate for an enzyme-catalyzed reaction, with substrate in excess, at 20.0°C is 2.52 μmol/min. At 30.0°C, the rate increases to 36.6 μmol/min. What is the activation energy, E_a, for this reaction? Hint: Use the Arrhenius equation, $k = e^{-E_a/RT}$, in its linear form,

$$\ln k = -E_a/R \times (1/T) + \ln A$$

where k is the rate constant for the reaction, A is the frequency factor, R is the gas constant (8.314 J/molK), and T is the kelvin temperature.

2. Figure 6.1 is the pH profile for the proteolytic enzyme pepsin, which operates in the stomach in a rather acidic environment. Pepsin, like the HIV-1 protease, is an aspartic protease, with two aspartate side chains in the active site. What information does Figure 6.1 provide in terms of the mechanism of the action of pepsin? What is a somewhat unusual aspect of this mechanism?

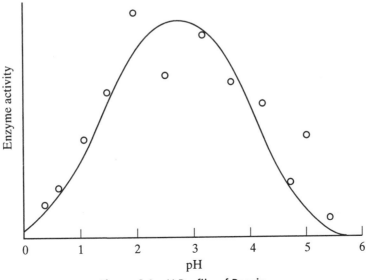

Figure 6.1 pH Profile of Pepsin.

3. Figure 6.2 shows an example of *burst kinetics* for an enzyme-catalyzed reaction of the type:

$$S \xrightarrow{\text{Enzyme}} P + Q$$

What mechanistic information is provided by this data?

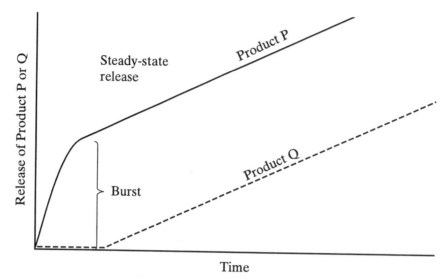

Figure 6.2 Burst Kinetics.

4. For the nonenzymatic reaction shown here, there are 2.50×10^{-2} mmol of product R formed in a two minute period using 0.0100 M concentrations of A and B.

$$A + B \rightarrow R + T$$

When an enzyme is used to catalyze the reaction, using the same concentrations of A and B (and all other conditions also kept the same), there are 4.00 mmol of product R formed in a two-minute period. Evaluate the proximity effect provided by the enzyme by computing the effective molarity (Section 6.6 in your text), assuming all rate enhancement is due only to the proximity effect.

5. Lysozyme catalyzes the hydrolysis of the polysaccharide component of bacterial cell walls, which consists of alternating N-acetylglucosamine (NAG) and N-acetylmuramic acid (NAM) residues. Lysozyme also accepts chitin as a substrate. Chitin is a polysaccharide in crustacean shells that contains only NAG residues connected by ß-1,4 linkages. The following data were collected in a study of lysozyme action on various oligomers of NAG.

Substrate	Relative Hydrolysis rate
NAG_2	0
NAG_3	1
NAG_4	8
NAG_5	4,000
NAG_6	30,000
NAG_8	30,000

What conclusions can be drawn about the substrate specificity of lysozyme? Which particular glycosidic bond of the substrate is cleaved by lysozyme?

7

Coenzymes

I. KEY TERMS

With the help of this study guide, your textbook, and class notes, you should be able to define and explain the significance of the following terms:

cofactor

holoenzyme

coenzyme

metal-activated enzyme

iron-sulfur cluster

prosthetic group

vitamin-derived coenzyme

apoenzyme

essential inorganic ion

reactive center

metalloenzyme

cosubstrate

metabolite coenzyme

group-transfer protein

II. EXERCISES

A. True–False

_____ 1. Iron-sulfur clusters are modified forms of heme groups involved in certain electron transfer processes.

_____ 2. Some coenzymes are made by an organism for its own use and are called metabolite coenzymes.

_____ 3. One should not take vitamin supplements that contain water-soluble vitamins, since an excess might lead to hypervitaminosis.

_____ 4. Pellagra is a disease characterized by dermatitis, impaired digestion, and diarrhea. Pellagra is associated with a deficiency of vitamin C.

_____ 5. Vitamin B_1, thiamine, deficiency is associated with the disease beriberi, which is characterized by rapid weight loss, muscle atrophy, and weakness.

_____ 6. Tetrahydrofolate, a coenzyme with a polyglutamate tail, is derived from the water-soluble vitamin folic acid.

_____ 7. Vitamin K is a lipid vitamin with antioxidant activity.

_____ 8. These reactants: $NADH + H^\oplus + Coenzyme\ Q$ are more likely to produce products than these reactants: $NAD^\oplus + Coenzyme\ QH_2$.

_____ 9. Cytochromes are heme-containing protein coenzymes in which the iron (III) ions of the heme undergo reversible one-electron reduction.

_____ 10. The four fat-soluble vitamins (A, D, E, and K) all function as coenzymes.

B. Short Answers

1. Cofactors can be classified in one of two major categories, which are _____ and _____ .

2. Some enzymes do not require cofactors to be active. For those that do, the inactive protein itself, without its cofactor, is called the _____ .

3. When the inactive protein is combined with its specific cofactor, an active form is created called the _____ .

4. The function of coenzymes is to act as _____ .

5. Mobile metabolic groups transferred by coenzymes are temporarily attached to the _____ of the coenzyme.

6. A coenzyme that is chemically altered during an enzyme-catalyzed reaction and that dissociates from the active site is called a/an _____ .

7. A coenzyme that does not dissociate from the active site but remains firmly bound to the enzyme is called a/an _____ .

8. A molecule with opposite ionic charges on adjacent atoms is called a/an _____ .

9. The two main categories of vitamins are the _____ vitamins and the _____ vitamins.

10. The _____ are proteins that act as coenzymes. They usually contain such reaction centers as _____ , _____ , and _____ .

C. Problems

1. Explain the difference between metal-activated enzymes and metalloenzymes.

2. Explain why coenzymes are required by some enzymes.

3. For each of the following coenzymes, indicate a typical function or role:

   Coenzyme A _____ .

$NAD^{\oplus}$ _____.

Biotin _____.

Thiamine pyrophosphate _____.

Tetrahydrofolate _____.

ATP _____.

FAD _____.

4. Which vitamin-derived coenzyme exists as a lactone? What is a lactone? What is the function of this coenzyme?

5. The reduced form of FAD is $FADH_2$. Some older texts also showed the reduced form of $NAD^{\oplus}$ as $NADH_2$. Discuss the currently-accepted mechanism for the reduction of $NAD^{\oplus}$ and how it does not justify the use of the representation $NADH_2$ for the reduced form of this coenzyme.

6. Describe the sequence of steps in the addition of substrate and cosubstrate and the formation of product(s) in the oxidation of pyruvate to lactate by lactate dehydrogenase. Is this a ping-pong mechanism?

7. A histidine and an arginine residue are known to play roles in the active site of lactate dehydrogenase. Explain their involvement.

8. Both $NAD^{\oplus}$ and FAD (also FMN) are coenzymes for dehydrogenase enzymes. They differ not only in the manner in which they are held by the enzymes, but also in the way that electrons are transferred. Explain the differences and show the mechanisms by which $NAD^{\oplus}$ and FAD are respectively reduced.

9. What mammalian coenzyme has a long hydrophobic chain of isoprenoid units and what is the function of this chain?

10. What is the reduction potential of a cytochrome and why does it vary from one cytochrome to another even if the cytochromes of a given class (*a*-cytochromes, *b*-cytochromes, *c*-cytochromes) have the same heme groups?

D. Additional Problems

1. Pyruvate carboxylase is an enzyme that catalyzes the carboxylation of pyruvate to form oxaloacetate. The enzyme requires biotin as a coenzyme. ATP is also needed. Write the mechanism for this enzymatic reaction.

2. How does inhibition of the enzyme dihydrofolate reductase help retard the growth of cancer cells?

3. In one of the "Rocky" movies, Sylvester Stallone drank a concoction that contained raw eggs. He saved time with respect to cooking, but what health problem might arise from consuming raw eggs too often? What coenzyme is involved?

4. (a) What coenzyme forms Schiff bases? (b) What are aldimines? (c) What is the difference between internal and external aldimines with respect to the mechanism of transamination?

5. Although somewhat rare, pernicious anemia may develop in some older people and in long-term strict vegetarians. (a) What coenzyme is involved/affected in pernicious anemia and, (b) what are some unusual aspects of this coenzyme? (c) What is possibly different in the development of this disease in older people as opposed to its development in the long-term vegetarians?

8

Carbohydrates

I. KEY TERMS

With the help of this study guide, your textbook, and class notes, you should be able to define and explain the significance of the following terms:

monosaccharide
polysaccharide
glucoconjugate
heteroglycan
aldose
epimer
furanose
anomer
alduronic acid
glycoside
aglycone

proteoglycan
peptidoglycan
glycoform
N-linked oligosaccharide
amylopectin
oligosaccharide
glycan
homoglycan
ketose
triose

pyranose
anomeric carbon
aldonic acid
glycosidic bond
reducing sugar
limit dextrin
glycosaminoglycan
glycoprotein
O-linked oligosaccharide
amylose

II. EXERCISES

A. True–False

_____ **1.** Dihydroxyacetone could be described as a ketotetrose.

_____ **2.** The naturally-occurring sugars, like the natural amino acids, have a stereochemistry that places them in the L-series of sugars.

_____ **3.** Epimers are two carbohydrates that differ only in the configuration of groups at a single chiral center.

_____ **4.** Monosaccharides that form six-membered rings are called pyranoses.

_____ **5.** The complete hydrolysis of starch, glycogen, and cellulose would yield only glucose.

_____ **6.** Amylopectin is an entirely linear polymer of D-glucose units found in natural starch.

_____ **7.** Lactose is a disaccharide that contains a β–(1→4) type linkage.

_____ **8.** Maltose, lactose, sucrose, and cellobiose are all reducing sugars.

_____ **9.** Glycogen is like the amylopectin of starch except that glycogen has β–(1→4) linkages between glucose units.

_____ **10.** Fructose forms a cyclic structure that is a hemiketal.

B. Short Answers

1. A carbohydrate that is made by joining five monosaccharides is called a/an _____.

2. The new chiral center formed when a monosaccharide such as galactose assumes a cyclic form is called the _____.

3. The two different forms of fructose that exist in solution, which differ in the –OH group orientation at the number 2 carbon, are called _____.

4. The non-sugar portion of a glycoside is called a/an _____.

5. The digestion of starch by amylase produces an amylase-resistant core called a/an _____.

6. Proteins with identical amino acid sequences but with different oligosaccharide chain compositions are special types of glycoproteins called _____.

7. The homoglycan _____, composed of _____ units in β–(1→4) type linkages, is found in the exoskeletons of insects and crustaceans.

8. Bacterial cell walls contain a heteroglycan which is classified as a/an _____.

9. Large glycoproteins, which contain _O_-linked oligosaccharides and are found in mucus, are called _____.

10. An epimer of D-fructose is _____.

C. Problems

1. Draw the Fischer projection of the enantiomer of D-galactose.

2. Draw the structure of 2-deoxy-α-D-ribofuranose.

3. If approximately 4% of the glucose units in amylopectin are involved in α–(1→6) linkages where branching occurs, how many branches would there be on a linear portion of amylopectin that had 212 glucose units in that linear portion?

4. Explain why sucrose is not a reducing sugar when it is composed of glucose and fructose, both of which are reducing sugars.

5. List the monomeric units that make up the following:

a. amylopectin _____

b. chitin _____

c. milk sugar _____

d. maltose _____

e. glycogen _____

6. Draw the structure of β-D-mannopyranosyl-(1→5)-α-D-ribofuranose. Is this a reducing sugar?

7. It is possible to reduce the aldehyde group of aldoses like glucose, mannose, and ribose to an alcohol group. These derivatives are called sugar alcohols. Some (xylitol) have been used in sugarless gums (Orbit™) and other similar products. Draw the structures of the sugar alcohols glucitol and ribitol. Will they exist in cyclic or open chain forms? Explain.

8. Assume that glycogen has a molar mass of about 3.0×10^6 g. (a) How many individual glucosyl residues does it contain? (b) How many of these are located at branch points?

9. If a total of about a pound of glycogen can be stored in the body of an adult male, how many molecules of the size indicated in the previous problem are present in this amount of glycogen?

10. (a) The linear Fischer projection form of an aldohexose has how many chiral carbons? (b) Based on this, how many isomers are possible? Compare your answer to the number of aldohexoses shown in text Figure 8.3. (c) What do you conclude?

11. Make the same comparison based on the Haworth projection form of an aldohexose. What do you conclude?

D. Additional Problems

1. Because it contains a number of chiral centers, glucose is optically active. It has been found, however, that a solution of the pure α-anomer has an initial "specific rotation" value different from that of a solution of the pure β-anomer. The initial value for α-D-glucose is +112.2° while that for β-D-glucose is +18.7°. In both cases, these values change with time (that for the α-anomer decreases and that for the β-anomer increases) until an equilibrium value is reached. Use the percentage of each anomer present in aqueous solution as indicated in your text and calculate the equilibrium value of the specific rotation for glucose.

2. The specific rotation, [α], of a compound is related to the observed optical rotation of a solution of that compound as indicated:

$$[\alpha] = A/lc$$

where A is the observed optical activity in degrees, l is the cell path length in decimeters, and c is the concentration in grams per milliliter. Sucrose is dextrorotatory with a specific rotation of about +66.5°. A sucrose solution measured in a 25-cm polarimeter tube showed an optical rotation of +46.8°. What is the concentration of the sucrose solution?

3. Some sources indicate that amylose forms a helix in which there are six glucosyl residues per turn of the helix and that each turn occupies a linear distance of 0.8 nm. (a) How long is a helical segment of amylose that contains 900 glucosyl residues? (b) How does this helix compare to the α-helix of proteins?

4. Reducing sugars, such as D-galactose, react with $Ag^{\oplus}$ in aqueous ammonia causing elemental silver to be deposited on the walls of the container. The reaction is known as the Tollen's silver mirror test. The open chain form of the sugar is required for a positive test. (a) Write the equation for this reaction with D-galactose and name the sugar derivative formed. (b) If less than 1% of the open chain form of the sugar is normally present in aqueous solution, how can this reaction occur at all?

5. Recent research suggests that glucose may be connected with the aging process as well as some of the complications associated with diabetes. Researchers think that the open chain form of glucose reacts with amine groups of proteins to first form a Schiff base which rearranges to an Amadori product (a keto derivative formed from a Schiff base by two tautomeric shifts of hydrogen atoms.) The Amadori product (an N-substituted 1-amino-1-deoxy-2-ketose) is thought to react with another glucose and an amine group of another protein to form a cross link between the protein molecules. This type of cross link has been identified as 2-furanyl-4(5)-(2-furanyl)-1H-imidazole (FFI). It is suspected that cross links of this type cause collagen to become tough and inflexible. Other proteins such as those in the lens of the eye may also be affected, leading to cataract formation. (a) Use appropriate structures to illustrate the formation of a Schiff base with an amino group of a protein, and its rearrangement to form an Amadori product. (b) Also illustrate the formation of the FFI cross link between two proteins.

   For a little help: All carbons in the FFI cross-link are supplied by the two glucose molecules, including the carbons of the two furanyl moieties and the imidazole moiety. One of the furanyl moieties is bridged to the imidazole ring by a carbonyl group, the carbon of which came from one of the glucose molecules. The two nitrogen atoms of the imidazole ring were supplied by the amino groups of the two protein chains which were cross linked.

6. How many different disaccharides can be made from D-glucopyranose and D-fructofuranose? (Remember that disaccharide linkages involve at least one anomeric carbon.)

9

Nucleotides

I. KEY TERMS

With the help of this study guide, your textbook, and class notes, you should be able to define and explain the significance of the following terms:

nucleoside

ribonucleotide

purine

kinase

phosphatase

alarmone

nucleotide

deoxyribonucleotide

pyrimidine

induced fit

second messenger

cyclic ADP-ribose

II. EXERCISES

A. True–False

_____ **1.** Nucleotides are molecules that contain only a sugar and a nitrogenous base.

_____ **2.** Nucleosides are obtained as breakdown products of nucleic acids and found to contain a pentose and a nitrogenous base held to the pentose by a β-N-glycosidic bond.

_____ **3.** Uracil and thymine are members of the pyrimidine family and are typically found as components of DNA.

_____ **4.** The antibiotic cordycepin inhibits bacterial RNA synthesis because it is so similar to the adenosine moiety of the AMP nucleotide used in RNA synthesis.

_____ **5.** Kinases are enzymes that catalyze the transfer of a phosphoryl group, usually from ATP, to another compound.

_____ **6.** The term induced fit refers to the bending of a substrate molecule to properly fit in the active site of an enzyme.

_____   **7.** Phosphatases are enzymes that belong to the hydrolase class of enzymes.

_____   **8.** Signal nucleotides act intracellularly to alter the activity of certain enzymes.

_____   **9.** The nucleoside moieties of nucleic acids exist predominately in the *anti* conformations.

_____   **10.** Nucleoside diphosphate kinase will accept only ADP and ATP as substrates.

B. Short Answers

1. 2,4-Dioxopyrimidine is more commonly known as _____.

2. 2-Oxo-4-aminopyrimidine is more commonly known as _____.

3. The _____ tautomeric forms of the bases guanine, thymine, and uracil are more stable and thus predominate in cells over the _____ tautomers.

4. Deoxythymidine 5′-triphosphate is commonly abbreviated as _____.

5. The common name of the enzyme that catalyzes the conversion of GMP to GDP is _____.

6. The compound that donates glucose in the synthesis of glycogen in animals is _____.

7. The compound that donates glucose in the synthesis of starch in plants is _____.

8. A signal nucleotide found in both invertebrate and mammalian cells, formed from $NAD^{\oplus}$ by displacement of the nicotinamide ring, is called _____.

9. Cyclic nucleotides such as cAMP and cGMP are formed as a result of some extracellular stimulus, and because they act to deliver that message within the cell, they are referred to as _____.

10. DNA polymerase requires _____ as substrates for the extension of a DNA chain.

C. Problems

1. Explain how the sponge nucleoside cytosine arabinoside combats cancer.

2. In what state of ionization does ATP normally exist in cells?

3. GDP can be formed from GMP by the action of guanylate kinase, which requires ATP as a second substrate. If the equilibrium constant for this reaction is about 1.0, how can the formation of GDP be spontaneous?

4. Adenylate kinase catalyzes the reaction

$$\text{AMP} + \text{ATP} \longleftrightarrow 2\,\text{ADP}$$

No ADP is formed until both AMP and ATP are bound to the enzyme. What does this suggest about the enzymatic mechanism? Is covalent catalysis likely involved?

5. What is the function of the enzyme nucleoside diphosphate kinase? What is (are) its substrate(s)?

6. When free glucose is degraded in the cell by the glycolytic pathway, it is converted to glucose 6-phosphate. On the other hand, in the reaction by which UDP-glucose is formed as shown in text Figure 9.17, α-D-glucose 1-phosphate is the substrate that supplies the glucose moiety. Why do you suppose α-D-glucose 1-phosphate might be a better precursor for UDP-glucose than glucose 6-phosphate?

7. When UDP-glucose is formed from α-D-glucose 1-phosphate and UTP as shown in text Figure 9.17, pyrophosphate, PP_i, is formed and is rapidly hydrolyzed to 2 P_i by the action of pyrophosphatase. Why is this hydrolysis step important?

8. Explain how an extracellular signal is communicated to an intracellular location in the regulation of cellular metabolism in vertebrates.

9. What are "alarmones"? Cite an example.

10. What is cyclic ADP-ribose? What is its function?

D. Additional Problems

1. One of the effects of cyclic AMP is the production of adrenaline. Caffeine, 1,3,7-trimethylxanthine, is a stimulant. Cyclic AMP is converted to the inactive 5′-AMP by the action of a phosphodiesterase. (a) What is xanthine? (b) Use this information to suggest a means by which caffeine acts as a stimulant.

2. The antitumor agent 5-fluorouracil becomes 5-fluoro-2′-deoxyuridine-5′-phosphate in the body, and is known to inhibit the enzyme thymidylate synthase, as indicated in the text, problem 3. Speculate on the type of inhibition exhibited by this agent and suggest how this inhibition serves to combat cancer.

3. Briefly discuss six metabolic functions performed by nucleotides and their derivatives.

4. The drug AZT (3′-azido-2′,3′-dideoxythymidine) mentioned in text problem 4., must be activated before it combats HIV infection. Speculate on what this activation might involve and on the general metabolic process(es) affected by AZT.

5. If one adds ATP to a concentrated aqueous solution of the enzyme nucleoside diphosphate kinase mentioned in problem C.5., ADP can shortly thereafter be detected in solution. If one then adds CDP to the solution, CTP can shortly thereafter be detected in the solution. Based on this information, suggest a mechanism for the nucleoside diphosphate kinase reaction.

10

Lipids and Membranes

I. KEY TERMS

With the help of this study guide, your textbook, and class notes, you should be able to define and explain the significance of the following terms:

lipid
polyunsaturated
glycerophospholipid
plasmalogen
sphingosine
sphingomyelin
ganglioside
sterol
eicosanoid
lateral diffusion
transverse diffusion
fluid mosaic model
integral membrane protein
lipid-anchored membrane protein
prenylated protein

channel
symport
passive transport
primary active transport
endocytosis
transducer
second messenger
G-protein
monounsaturated
triacylglycerol
phosphatidate
sphingolipid
ceramide
cerebroside
steroid

wax
prostaglandin
lipid bilayer
flippase
freeze-fracture electron microscopy
peripheral membrane protein
pore
uniport
antiport
active transport
secondary active transport
exocytosis
effector enzyme
cascade
autophosphorylation

II. EXERCISES

A. True–False

_____ **1.** Phosphatidyl choline (also called lecithin) is a type of phospholipid found in biological membranes.

_____ **2.** Of the three types of membrane lipids mentioned in the text, all are charged.

_____ **3.** Plasmalogens are sphingosine derivatives.

_____ **4.** All phospholipases catalyze the specific hydrolysis of phosphoester bonds of phospholipids.

_____ **5.** The fatty acid component of a ceramide is attached to sphingosine via an ester bond.

_____ **6.** Gangliosides contain the *N*–acetylneuraminic acid group.

_____ **7.** Micelles are aggregates of cholesterol.

_____ **8.** Membrane lipid bilayers are permeable to water.

_____ **9.** Triacylglycerols are neutral lipids.

_____ **10.** Membrane proteins are associated only with the surfaces of biological membranes.

_____ **11.** The two surfaces of a biological membrane differ from each other.

_____ **12.** Integral proteins can move within the lipid bilayer of the membrane.

B. Short Answers

1. Three major types of lipids in animal membranes are _____, _____, and _____.

2. A moiety common in the head group of all phosphatidates is _____.

3. Three common aminoalcohols found in phosphatidates are _____, _____, and _____.

4. _____ are glycosphingolipids that provide cells with distinctive surface markers such as those that provide the basis for ABO blood-grouping.

5. _____ are glycosphingolipids abundant in nerve tissue.

6. The type of enzyme that catalyzes formation of phosphatidates from glycerophospholipids is _____.

7. _____ contain sphingosine, a fatty acid, and a carbohydrate such as glucose, galactose, or an oligosaccharide.

8. _____ is an ATP-dependent protein that moves phospholipids from the outer to the inner leaflet of a membrane.

9. The more fluid phase of membranes is called the _____, and the less fluid phase is called the _____.

10. The movement of a phospholipid from one monolayer to the other in a membrane is called _____.

11. The two types of proteins associated with membranes are the _____ and _____ proteins.

C. Problems

1. List the products of the complete hydrolysis of the following: (a) glucocerebroside, (b) sphingomyelin, (c) lecithin.

2. How do cholesterol molecules orient themselves in biological membranes?

3. What are eicosanoids? Name three different types of eicosanoids that exhibit different biological functions.

4. What is the driving force for the formation of lipid mono– and bilayers in/on an aqueous solution?

5. How are carbohydrates arranged in membranes?

6. Why does treatment of biological membranes with chelating agents, such as ethylenediaminetetraacetic acid (EDTA), cause the release of peripheral proteins?

7. Explain the difference between primary and secondary active transport in membranes.

8. Explain how integral proteins are associated with the lipid bilayer.

9. Thin layer chromatography of a triacylglycerol and lecithin on silica gel in chloroform–methanol– water developing solvent shows that lecithin has an R_f value of about 0.4 and the triacylglycerol travels nearly with the solvent front. Why are they so different?

10. Apparently the fluidity of membranes is important to their biological functions. It was shown that in *E. coli* the ratio of saturated to unsaturated fatty acids decreases as the temperature decreases. How does the presence of unsaturated fatty acids affect the fluidity of membranes? What role does cholesterol play in mammalian membrane fluidity?

11. Briefly contrast the mode of lipid vesicle formation occurring in endocytosis with that occurring in exocytosis.

D. Additional Problems

1. If there are 2.86×10^4 phospholipid molecules in a section of lipid bilayer that has a surface area on each face of the membrane of 100.0 μm^2, what is the surface area occupied by a single molecule?

2. A liposome is a vesicle formed by dispersion of phospholipids in aqueous salt solutions. If such a liposome had a diameter of 40.0 nm, what is the volume of aqueous solution enclosed by the liposome? The thickness of a lipid bilayer is about 4–5 nm.

3. Studies have indicated that the diffusion constant D for a phospholipid in a membrane is about 10^{-8} cm^2/sec. The distance s (in cm) traveled by a molecule in t seconds is given by: $s = (4Dt)^{1/2}$. How long would it take for a lipid molecule to travel the length of a bacterium that is about 2.0 μm in length?

4. Membrane asymmetry is often studied by the use of 2,4,6–trinitrobenzene sulfonic acid (TNBS), which undergoes nucleophilic attack by the primary amine groups of certain lipids. What specific membrane lipids will react with TNBS? Write a generalized equation (just show the functional group on a "stick" lipid) for this reaction. Hint: Sulfite is liberated. For an additional challenge, write the mechanism for this reaction.

5. Bacteriorhodopsin is the only type of protein in the "purple membrane" of *Halobacterium halobium*. The membrane is about 4 nm thick and the protein contains 247 amino acid residues. Portions of the protein which traverse the membrane are arranged in the α–helix type of secondary structure. Only 27% of the amino acid residues of the protein are not involved in these α–helical areas. How many times does the bacteriorhodopsin chain traverse the membrane?

11

Introduction to Metabolism

I. KEY TERMS

With the help of this study guide, your textbook, and class notes, you should be able to define and explain the significance of the following terms:

metabolism	metabolically irreversible reaction	feed-forward activation
catabolic reaction	energy-rich compound	protein phosphatase
pathway	phosphoryl-group-transfer potential	citric acid cycle
feedback inhibition	oxidizing agent	standard state
protein kinase	electromotive force	near-equilibrium reaction
glycolysis	metabolite	phosphagen
oxidative phosphorylation	anabolic reaction	reducing agent
mass-action ratio	flux	metabolic fuel

II. EXERCISES

A. True–False

_____ **1.** Metabolism is an inclusive term used to indicate all reactions carried out by living cells.

_____ **2.** The enzymes of the glycolytic pathway are found in the cell cytoplasm. Since they are not compartmentalized, glycolysis is always "on" and will convert glucose to pyruvate as long as any glucose is available.

_____ **3.** Cells are very efficient at capturing the stored energy in a molecule such as glucose; thus, the efficiency of energy conversion approaches 100%.

_____ **4.** One gram of NaCl dissolved in a test tube of water has a greater entropy than does a gram of NaCl in the form of a crystalline solid.

_____ **5.** For a reaction to be spontaneous, the change in free energy, ΔG, must be negative.

_____ **6.** Processes that result in an increase of the entropy of the system are always spontaneous.

_____ **7.** The free energy change, ΔG, in a system is dictated only by the absolute temperature and the change in heat content (enthalpy), ΔH, of the system.

_____ **8.** Cellular reactions tend to reach equilibrium so that ΔG becomes zero.

_____ **9.** Cellular reactions for which the steady state ratio of products to reactants, Q, is numerically not close to the normal equilibrium constant, K_{eq}, are known as metabolically irreversible reactions.

_____ **10.** For a reaction of the type $A + B \rightleftharpoons C + D$, the value of the equilibrium constant ($K_{eq} = 0.125$) indicates that the reaction proceeds nearly to completion.

_____ **11.** In biological systems, ATP plays a central role in energy transfer since it has an intermediate phosphoryl transfer potential.

B. Short Answers

1. _____ reactions are those by which the cell synthesizes needed materials such as proteins and nucleic acids, while _____ reactions are those by which the cell degrades materials for energy and building blocks.

2. Three "forms" of metabolic pathways are _____, _____, and _____.

3. Two events that increase the likelihood of a spontaneous process are a large _____ in the entropy of the system and a/the _____ of heat by the system.

4. The multistep catabolism of glucose to pyruvate is called _____.

5. The type of bond in ATP or ADP that has the highest free energy of hydrolysis is called a/an _____.

6. A molecule or species that loses one or more electrons in a reaction is called a/an _____ agent and becomes _____.

7. Energy-rich phosphate-storage molecules that occur in muscle tissue are called _____. Two examples are _____ and _____.

8. Phosphorylated compounds with phosphoryl group-transfer-potentials equal to or higher than that of ATP are termed _____.

9. Metabolic pathways are controlled by the modulation of allosteric enzymes. The rate of the pathway can be slowed by the action of a/an _____ of the pathway, a process called _____. The pathway rate can be increased by the action of a/an _____, a process called _____.

10. The relative tendency, in an oxidation-reduction reaction, for one species to accept electrons from another species (be reduced) is termed the _____.

C. Problems

1. Can a reaction with a positive standard free energy change, $\Delta G^{0'}$, proceed in the forward direction? Explain.

2. For a reaction with a positive $\Delta G^{0'}$ value, what factor(s) would yield a negative ΔG such that the reaction would proceed spontaneously in the forward direction?

3. Using the information in text Table 11.3, determine the equilibrium constant, K_{eq}, for the following reaction at 25°C:

$$\text{Glucose 1-phosphate} + H_2O \rightarrow \text{Glucose} + P_i$$

4. In addition to regulation of pathways by allosteric control, pathway flux may also be controlled by covalent modification of interconvertible enzymes, such as by the addition or removal of phosphoryl groups. Are catabolic enzymes activated or deactivated by phosphorylation? Are anabolic enzymes activated or deactivated by phosphorylation?

5. Cite two or three consequences of the fact that various metabolic processes are sequestered in specialized subcellular compartments.

6. The in vitro oxidation of glucose to CO_2 and H_2O proceeds with a $\Delta G^{0'}$ of –686 kcal/mol. The in vivo oxidation of glucose results in the formation of 32 moles of ATP from ADP and P_i. (See text Section 13.5.) Determine the percentage of the "available" energy in glucose that is captured as ATP by cellular metabolism. The value for the total moles of ATP, 32, is obtained by tabulating all substrate level phosphorylation in glycolysis and the citric acid cycle as well as that formed by oxidation of all reduced cofactors formed (NADH and QH_2), via the electron transport system.

7. The hydrolysis of glucose 6-phosphate to glucose and P_i has a $\Delta G^{0'}$ value of –14 kJ/mol. If the steady state concentrations in a certain tissue of glucose 6-phosphate, glucose, and P_i were 10^{-3} M, 2×10^{-4} M, and 5×10^{-2} M, respectively, what is the change in free energy, ΔG, for the hydrolysis at normal body temperature of 37°C?

8. Imagine, in a certain biochemical pathway, that the standard free energy, $\Delta G^{0'}$, for the formation of a thioester, R-CO-SR′, from the reagents R-COOH and R′-SH is 10. kJ/mol. (a) Use the information below to propose a two-step process by which the cell might spontaneously form the thioester. (b) Calculate the actual free energy change for the two-step process. Should this process proceed spontaneously? Explain.

$$\text{ATP} + H_2O \rightarrow \text{AMP} + PP_i \qquad \Delta G^{0'} = -32 \text{ kJ/mol}$$

9. Your text illustrated the calculation of the standard free energy change, $\Delta G^{0'}$, for the oxidation of NADH by the electron transport chain with oxygen as the ultimate acceptor of electrons. Do the same kind of calculation for the oxidation of succinate by Q (ubiquinone).

10. The first reaction of the citric acid cycle, the condensation of oxaloacetate with acetyl CoA, has a standard free energy change of –32.2 kJ/mol. The last reaction in the cycle, regeneration of oxaloacetate from malate, has a standard free energy change of 29.2 kJ/mol. Explain how the cycle can continue to operate when the reaction that regenerates oxaloacetate has a positive standard free energy change.

D. Additional Problems

1. Fully protonated ATP could be written as H_4ATP. However, ATP is extensively ionized at physiological pH. Consider this equilibrium:

$$HATP^{3\ominus} \rightleftharpoons H^{\oplus} + ATP^{4\ominus} \qquad pK_a = 6.95$$

(a) Barring any other influences, what would be the percentage of the fully ionized form of ATP in blood of normal pH 7.4? (b) How does this differ from the percentage of fully ionized ATP in cells of pH 7.0?

2. In the glycolytic pathway, phosphorylation of free glucose is catalyzed by hexokinase with ATP as the phosphoryl-group donor. (a) If the $\Delta G^{0'}$ is -16.7 kJ/mol for this reaction, what is the value of the equilibrium constant at 37°C? (b) If the initial concentrations of glucose and ATP were each 0.0500 M, what would be the equilibrium concentrations of all species? (Note: This hypothetical problem may not represent actual cellular concentrations of glucose and ATP and also supposes that an equilibrium could be reached. Actually, the product glucose 6-phosphate is immediately used in the next reaction of the pathway to maintain a steady state of flux through the pathway.)

3. In erythrocytes, the steady state concentration of ATP is estimated to be 13 times that of ADP. The standard free energy change, $\Delta G^{0'}$, for glucose phosphorylation catalyzed by hexokinase is -16.7 kJ/mol, while the actual free energy change, ΔG, is -33.9 kJ/mol. Under these conditions, what would be the ratio of glucose to glucose 6-phosphate at 37°C?

4. In a cell at 37°C, the actual free energy change, ΔG, for the hydrolysis of ATP to ADP and P_i was determined to be -41.8 kJ/mol. Consider the standard free energy change, $\Delta G^{0'}$, given for this reaction in text Table 11.1 and estimate the ratio of ATP to ADP in this cell. Also indicate whether the hydrolysis of ATP is spontaneous under these conditions.

5. Some microorganisms form lactate from pyruvate (the product of glycolysis); this also occurs in exercising muscle. One source gives the standard free energy change for the net conversion of glucose to lactate as -123.5 kJ/mol. The overall reaction is:

$$\text{Glucose} + 2\ \text{ADP} + 2\ P_i \rightarrow 2\ \text{lactate} + 2\ \text{ATP}$$

The normal glycolytic conversion of glucose to pyruvate has a standard free energy change of about -73 kJ/mol according to this net reaction:

$$\text{Glucose} + 2\ \text{ADP} + 2\ P_i + 2\ \text{NAD}^{\oplus} \rightarrow 2\ \text{pyruvate} + 2\ \text{ATP} + 2\ \text{NADH} + 2\ \text{H}^{\oplus}$$

Contrast these two systems in terms of (a) energy production, (b) extent of oxidation of glucose, and (c) relative likelihood of the reactions to proceed spontaneously as written.

12

Glycolysis

I. KEY TERMS

With the help of this study guide, your textbook, and class notes, you should be able to define and explain the significance of the following terms:

glycolysis   isozyme   Pasteur effect

substrate-level phosphorylation

II. EXERCISES

A. True–False

_____ **1.** Glycolytic enzymes are found in mitochondria.

_____ **2.** Overall, glycolysis is an energy-releasing (exergonic) process.

_____ **3.** Glycolysis involves the anaerobic oxidation of glucose.

_____ **4.** Alcoholic fermentation in yeast duplicates the glycolytic process except for two final steps.

_____ **5.** Humans have all the enzymes but one necessary for alcoholic fermentation.

_____ **6.** Reaction eight of glycolysis, the conversion of 2-phosphoglycerate to 3-phosphoglycerate, is catalyzed by phosphoglycerate mutase. Since the product of the reaction is an isomer of the substrate, the enzyme is classified as an isomerase.

_____ **7.** Glycolysis consists of two stages, the hexose stage and the triose stage. A difference is that the hexose stage requires ATP and the triose stage produces ATP.

_____ **8.** The third reaction of glycolysis is regulated in addition to the first reaction of glycolysis, since substrates can enter the pathway as fructose 6-phosphate, bypassing the first reaction.

_____ **9.** In mammals, free glucose enters cells under the control of insulin and with the aid of passive transporters called GLUT 1, 2, 3, and so on. Glucose remains in its free form until ATP levels are low and it enters the glycolytic pathway.

_____ **10.** Fructose, mannose, and galactose are all catabolized by glycolysis after conversion to appropriate intermediates of the pathway. The moles of ATP produced by glycolysis is the same for all of these hexoses.

B. Short Answers

1. Pyruvate kinase occurs as _____, different forms of the enzyme that catalyze the same reaction, but may have different K_m values for the substrate, _____.

2. Enzymes which catalyze the hydrolytic removal of a phosphoryl group from a substrate are called _____.

3. Pyruvate kinase isozymes are allosterically activated by _____, and inhibited by _____.

4. Three glycolytic enzymes under metabolic control are _____, _____, and _____

5. The enzyme-catalyzed formation of ATP or some other high-energy nucleotide directly in a pathway step is called _____.

6. An enzyme that catalyzes the conversion of one sugar phosphate into another by changing the orientation of a single –OH group is called a/an _____.

7. The slowing of glycolysis in the presence of oxygen is called _____

8. Lactose intolerance, a condition prevalent in adults in most parts of the world, is due to _____.

9. Two negative allosteric modulators of the enzyme phosphofructokinase-1 are _____ and _____

10. The product(s) of the glycolytic step catalyzed by pyruvate kinase is(are) _____.

C. Problems

1. Use Figure 12.1 to write the net reaction for the glycolytic formation of pyruvate from glucose. Structures are not necessary.

2. Assuming you had in solution all the reactants of the net reaction written in problem C.1. and all the glycolytic enzymes, would glycolysis proceed in vitro (in a test tube)?

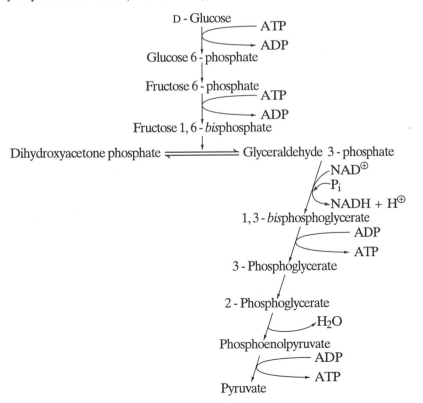

Figure 12.1 The Glycolytic Pathway.

3. Is lactate constantly formed in mammalian cells? Explain.

4. Fill in the blanks to show the initial glycolytic intermediates formed when each hexose is catabolized. Also indicate the number of moles of each intermediate formed per mole of each hexose.

Hexose	Initial Glycolytic Intermediate	Number of moles of Intermediate
mannose	_____	_____
fructose	_____	_____
galactose	_____	_____

5. What is the importance of the function of the enzyme triose phosphate isomerase in glycolysis?

6. Illustrate the cyclic process by which $NAD^{\oplus}$ is regenerated to keep glycolysis going. Not all intermediates need be shown. How is $NAD^{\oplus}$ regeneration achieved in fermentation?

7. During glycolysis, what intermediate(s) accumulate(s) if an enolase inhibitor is present?

8. What would be the effect in an individual in whom the enzyme triose phosphate isomerase was either deficient or inhibited?

9. Iodoacetate inhibits glyceraldehyde 3-phosphate dehydrogenase by interacting with an active site cysteine residue. Show the reaction that occurs between iodoacetate and the enzyme. What kind of inhibition is this?

10. In the triose stage of glycolysis, ATP is formed by the transfer of a phosphoryl group from a glycolytic intermediate to ADP. In what forms do these phosphoryl groups enter the pathway?

D. Additional Problems

1. Use Table 12.1 to determine the standard free energy change, $\Delta G^{0'}$, for the glycolytic conversion of one mole of glucose to two moles of pyruvate.

Table 12·1 Standard Free Energy Changes for Glycolysis

Step No.	Reaction	Enzyme	$\Delta G^{0'}$, kJ/mol
1	Glucose → G6P	hexokinase	−16.7
2	G6P → F6P	glucose 6-phosphate isomerase	+1.67
3	F6P → F1,6bisP	phosphofructokinase–1	−14.2
4	F1,6bisP → DHAP + G3P	aldolase	+24.0
5	DHAP → G3P	triose phosphate isomerase	+7.66
6	G3P → 1,3bisPG	glyceraldehyde 3–phosphate dehydrogenase	+6.28
7	1,3bisPG → 3PG	phosphoglycerate kinase	−18.8
8	3PG → 2PG	phosphoglycerate mutase	+4.44
9	2PG → PEP	enolase	+1.84
10	PEP → pyruvate	pyruvate kinase	−31.4

2. From the standpoint of the energy available from conversion of glucose to pyruvate by glycolysis, determine the percent efficiency of glycolysis based on the energy captured as ATP.

3. In exercising muscle, some of the glycolytic product pyruvate is converted to lactate. Is the efficiency of glycolysis increased or decreased by this additional step? Determine the percent efficiency for the production of two moles of lactate from one mole of glucose by glycolysis. The conversion of pyruvate to lactate is accompanied by a $\Delta G^{0'}$ of −25.1 kJ/mol.

4. The standard free energy change, $\Delta G^{0'}$, for the cleavage of fructose 1,6-bisphosphate into dihydroxyacetone phosphate and D-glyceraldehyde 3-phosphate (step 4) is +24 kJ/mol. With this large positive standard free energy change, how can glycolysis continue?

5. An enolase deficiency has an adverse effect on erythrocytes, specifically with respect to the delivery of oxygen to the tissues. Explain.

13

The Citric Acid Cycle

I. KEY TERMS

With the help of this study guide, your textbook, and class notes, you should be able to define and explain the significance of the following terms:

citric acid cycle

porin

mobile carrier of reducing power

prochiral molecule

glyoxylate cycle

amphibolic

pyruvate translocase

oxidative phosphorylation

anaplerotic

II. EXERCISES

A. True–False

_____ **1.** FAD is used in the citric acid cycle as an oxidizing agent.

_____ **2.** The final steps in the oxidation of glucose to CO_2 occur in the citric acid cycle.

_____ **3.** Massive amounts of all the intermediates of the citric acid cycle must be present for its efficient operation.

_____ **4.** There are more six-carbon intermediates in the citric acid cycle than four-carbon intermediates.

_____ **5.** In eukaryotes, the enzymes that catalyze the reactions of the citric acid cycle are found in the cytosol.

_____ **6.** Intermediates of the citric acid cycle that are used for anabolic (synthetic) needs of the cell are replenished by anaplerotic reactions.

_____ **7.** Porphyrin biosynthesis could potentially interfere with the citric acid cycle by depleting succinyl CoA.

_____ **8.** Two moles of coenzyme A are consumed per mole of pyruvate catabolized to acetyl CoA that is subsequently consumed in the citric acid cycle.

_____ **9.** The citric acid cycle is controlled entirely by the availability of acetyl CoA and cycle intermediates.

_____ **10.** Per mole of glucose, the citric acid cycle produces as many high-energy phosphate molecules by substrate level phosphorylation as does glycolytic metabolism of glucose to pyruvate.

_____ **11.** A prochiral molecule is a symmetrical molecule that can be converted to a chiral molecule by substitution of a single functional group.

B. Short Answers

1. The oxidation of pyruvate to acetyl CoA and CO_2 requires the cofactors _____, _____, _____, _____, and _____.

2. Four dicarboxylic acid intermediates of the citric acid cycle are _____, _____, _____, and _____.

3. The conversion of isocitrate to α-ketoglutarate also produces _____. This is an oxidative process in which the actual oxidant is _____.

4. Transformation of pyruvate to acetyl CoA is catalyzed by _____.

5. The enzyme that catalyzes the transformation of α-ketoglutarate to succinyl CoA is _____.

6. The only step in the citric acid cycle that directly produces an energy-rich phosphoanhydride involves the conversion of _____ to _____.

7. There are a total of _____ oxidation-reduction steps in the catabolism of pyruvate via the citric acid cycle. The specific reduced coenzymes formed are _____ and _____.

8. The competitive inhibitor _____ slows the citric acid cycle and causes succinate to accumulate.

9. Certain citric acid cycle intermediates serve as biosynthetic precursors of other materials. Two such intermediates are _____ and _____.

10. A negative modulator involved in regulating each of the enzymes that catalyze the three metabolically irreversible reactions of the citric acid cycle is _____.

11. The intermediate _____ is formed by the hydration of _____ in the cycle.

C. Problems

1. What is coenzyme A and what is its function in the oxidation of pyruvate via the citric acid cycle?

2. List three sources of the acetyl group of acetyl CoA, a substrate of the citric acid cycle.

3. In eukaryotes, pyruvate is produced from glucose by glycolysis in the cytosol, but the enzymes of the citric acid are in the mitochondrial matrix (except for succinate dehydrogenase, which is part of Complex II in the inner mitochondrial membrane). How are these catabolic pathways connected despite being located in different cell compartments?

4. How does the process of covalent modification control the pyruvate dehydrogenase complex?

5. Explain the effect of each of the following on the citric acid cycle in mammals and tell how that effect is achieved.

   (a) an abundance of ADP

   (b) a low supply of oxaloacetate

6. Determine the number of moles of each reduced cofactor and energy-rich phosphoanhydride (ATP and/or GTP) formed from the following as each is catabolized, in mammalian cells, through the stage of CO_2 generation via the citric acid cycle:

   (a) one mole of pyruvate

   (b) one mole of acetyl CoA

   (c) one mole of glucose

7. Write the overall (net) reaction for the conversion of isocitrate to succinate, which is accomplished by means of several steps of the citric acid cycle. Use appropriate structures for isocitrate and succinate.

8. If minced pigeon breast muscle is treated with malonic acid and oxaloacetate, what is the effect on the citric acid cycle? Would a specific cycle intermediate accumulate?

9. Aconitase catalyzes the conversion of citrate to isocitrate with *cis*-aconitate as an intermediate. At equilibrium in vitro, the relative concentrations are 90% citrate, 4% *cis*-aconitate, and 6% isocitrate. With so little isocitrate formed, how can the citric acid cycle function adequately?

10. Acetyl CoA modulates the activity of the pyruvate dehydrogenase complex. How does this allow the cell to respond to varying levels of acetyl CoA?

D. Additional Problems

1. A person was diagnosed as having a hereditary pyruvate dehydrogenase deficiency. How could the citric acid cycle operate adequately under such a circumstance? Should carbohydrate be restricted in this person's diet? Explain.

2. Some biochemistry texts state that glucose cannot be synthesized from fatty acids. A more definitive statement in one text is "there can be no net synthesis of glucose from fatty acids, at least from the majority of fatty acids likely to be a part of the average diet." Oxaloacetate, in addition to being used for condensation with acetyl CoA to form citrate, can lead to glucose formation via gluconeogenesis. It has been demonstrated that, when labeled acetyl CoA from fatty acids is used in a functioning citric acid cycle, the label can be found in glucose. Explain how this can occur and yet be consistent with the "no net synthesis" statement above.

3. A derivative of pyruvate, hydroxypyruvate, must be converted to pyruvate for further catabolism. This seemingly simple change requires five separate reaction steps. The intermediates shown below are involved, as well as $NAD^{\oplus}$, NADH, ATP, and ADP. Show the five reaction steps in a pathway using $NAD^{\oplus}$, NADH, ATP, and ADP as needed. Hint: The last three steps of the sequence parallel those of glycolysis.

$$HO-CH_2-\underset{\underset{O}{\|}}{C}-COO^{\ominus}$$

hydroxypyruvate

Intermediates, not in order:

$$\begin{array}{cccc}
\text{COO}^{\ominus} & \text{COO}^{\ominus} & \text{COO}^{\ominus} & \text{COO}^{\ominus} \\
| & | & | & | \\
\text{CHOH} & \text{CHO}-\text{\textcircled{P}} & \text{CHOH} & \text{C}-\text{O}-\text{\textcircled{P}} \\
| & | & | & \| \\
\text{CH}_2\text{O}-\text{\textcircled{P}} & \text{CH}_2\text{OH} & \text{CH}_2\text{OH} & \text{CH}_2 \\
\\
\textbf{A} & \textbf{B} & \textbf{C} & \textbf{D}
\end{array}$$

4. A hereditary disorder called encephalomyelopathy (Leigh's disease) results in accumulation of pyruvate and lactate in blood. This disorder is due to a deficiency of the enzyme pyruvate carboxylase. Speculate on the effect this deficiency might have on the capacity of the individual to perform continuous work or exercise.

5. The equilibrium constant for the formation of citrate catalyzed by citrate synthase is 5×10^5.

$$\text{oxaloacetate} + \text{acetyl CoA} \xrightarrow{\text{citrate synthase}} \text{citrate} + \text{CoASH}$$

(a) Calculate the change in standard free energy, $\Delta G^{0\prime}$, for this reaction at 25°C.

Citrate can be transported from the mitochondrion into the cytosol and can be converted to acetyl CoA and oxaloacetate by the enzyme citrate lyase, according to this reaction:

$$\text{citrate} + \text{CoASH} + \text{ATP} \xrightarrow{\text{citrate lyase}} \text{oxaloacetate} + \text{acetyl CoA} + \text{ADP} + \text{P}_i$$

(b) Using information from text Table 11.1, calculate the change in standard free energy, $\Delta G^{0\prime}$, and the K_{eq} for this reaction at 25°C.

(c) Why is ATP required for this reaction? Is this requirement in harmony with the efficient operation of the citric acid cycle?

Additional Pathways In Carbohydrate Metabolism

I. KEY TERMS

With the help of this study guide, your textbook, and class notes, you should be able to define and explain the significance of the following terms:

glycogenolysis	glucose-alanine cycle	glucagon
limit dextrin	pentose phosphate pathway	gluconeogenesis
insulin	phosphorolysis	Cori cycle
epinephrine	glycogenin	substrate cycle
hormonal induction		

II. EXERCISES

A. True–False

_____ 1. In contrast to glycolysis, the pentose phosphate pathway allows the complete oxidation of glucose to CO_2.

_____ 2. The formation of DNA and RNA directly depends on high gluconeogenesis activity in the cell.

_____ 3. Glucose 1-phosphate is the direct product of glycogenolysis.

_____ 4. Glycogen synthase catalyzes the linkage of all glucose molecules used in the formation of glycogen in liver and muscle tissue.

_____ 5. Gluconeogenesis is simply a reversal of glycolysis that occurs when blood glucose levels fall below normal.

_____ **6.** The formation of one mole of glucose by gluconeogenesis from pyruvate requires the same amount of energy as that produced by glycolytic degradation of one mole of glucose to pyruvate.

_____ **7.** The Cori cycle is a combination of glycolysis and gluconeogenesis.

_____ **8.** The enzymes that catalyze the reactions of the pentose phosphate pathway are all found in the cytosol.

_____ **9.** NADPH is produced in the nonoxidative stage of the pentose phosphate pathway.

_____ **10.** Normally, the brain relies almost entirely on glucose for its energy needs.

B. Short Answers

1. The coenzyme required for reductive biosynthesis (eg. of fatty acids), that is produced by the pentose phosphate pathway, is _____.

2. A pentose phosphate pathway enzyme that catalyzes the transfer of a three-carbon unit from a ketose-phosphate to an aldose-phosphate is called a/an _____.

3. Name the pathway or process discussed in text Chapter 14 to which each of the following belongs.

pyruvate carboxylase _____   sedoheptulose 7-phosphate _____

glucose 6-phosphatase _____   UDP-glucose _____

4. Glycogen phosphorylase catalyzes the degradation of glycogen chains from their nonreducing ends but stops four glucose residues from a branch point. The remaining molecule is called a/an _____.

5. Gluconeogenesis requires four enzymes that are not enzymes of the glycolytic pathway. These four enzymes are _____, _____, _____, and _____.

6. The protein that is attached to the glycogen primer required for glycogen synthesis is called _____.

7. The principal hormones that control glycogen metabolism in mammals are _____, _____, and _____.

8. _____ is the process by which synthesis of PEP carboxykinase is increased due to increased transcription of the gene for this enzyme, triggered by increased levels of cAMP that result from prolonged release of glucagon.

9. Three major gluconeogenic precursors are _____, _____, and _____.

10. A pair of reactions that both form and degrade a specific substrate, in order to fine tune regulation of metabolism, is called a _____.

C. Problems

1. The liver is perfused in series with visceral tissues, while most other tissues are perfused in parallel. Of what consequence is this arrangement?

2. Figure 14.1 is the plot of kinetic data for the enzyme glycogen phosphorylase in the presence and absence of AMP. What can you conclude about this enzyme and the influence of AMP?

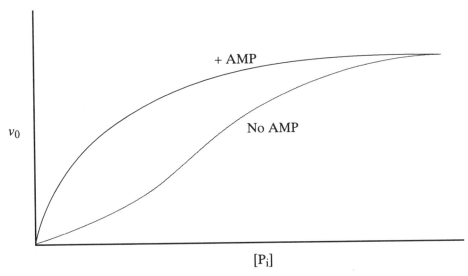

Figure 14.1 Effect of AMP on Glycogen Phosphorylase, with Constant Glycogen Concentration.

3. The gluconeogenic pathway uses four enzymes to bypass three reactions of glycolysis. Why are these bypass reactions necessary?

4. Contrast glycogenolysis in muscle and liver.

5. Glycogen is a highly branched polymer of glucose residues. Briefly explain how branches are formed in glycogen.

6. Your text says that "The level of PEP carboxykinase activity in cells limits the rate of gluconeogenesis." (a) Using appropriate structures, write the equation for the reaction catalyzed by PEP carboxykinase. (b) How is the level of PEP carboxykinase controlled?

7. (a) What is the Cori cycle? (b) What is its function?

8. (a) In what tissues is the pentose phosphate pathway usually most active?

   (b) What is the cellular location of the pathway?

   (c) What are the principal functions of the pathway?

9. (a) What reaction is catalyzed by the enzyme 6-phosphogluconate dehydrogenase?

   (b) In what pathway does this reaction occur?

10. Write the names of the expected products in each of the following.

   (a) Glyceraldehyde 3-phosphate + sedoheptulose 7-phosphate $\xrightarrow{\text{transketolase}}$

   (b) Fructose 6-phosphate + glyceraldehyde 3-phosphate $\xrightarrow{\text{transketolase}}$

   (c) Erythrose 4-phosphate + fructose 6-phosphate $\xrightarrow{\text{transaldolase}}$

D. Additional Problems

1. (a) Using the equation for the net reaction of gluconeogenesis shown in text Section 14.4, the information in Table 12.1 of this study guide, and the information below, calculate the standard free energy change, $\Delta G^{0'}$, for gluconeogenesis. (b) Is gluconeogenesis spontaneous under standard conditions? Explain.

   Given: Gluconeogenic conversion of pyruvate to phosphoenolpyruvate, catalyzed by pyruvate carboxylase and PEP carboxykinase: $\Delta G^{0'} = +2.1$ kJ/mol.

   Fructose 1,6-bisphosphate conversion to fructose 6-phosphate, catalyzed by fructose 1,6-bisphosphatase: $\Delta G^{0'} = -16.7$ kJ/mol.

   Glucose 6-phosphate conversion to glucose, catalyzed by glucose 6-phosphatase: $\Delta G^{0'} = -13.8$ kJ/mol.

2. Would gluconeogenesis be spontaneous under standard conditions if it proceeded by a reversal of all the reactions of glycolysis?

3. Glutathione (GSH), γ-Glu-Cys-Gly, protects red blood cells by reacting with peroxides that can cause degradation
   $\qquad\qquad\qquad\quad |$
   $\qquad\qquad\qquad\quad$ SH
   of fatty acids in cell membranes, and by the reduction of methemoglobin. In so doing, glutathione becomes oxidized to glutathione disulfide (GSSG). Glutathione is regenerated by the action of the enzyme glutathione reductase, which requires NADPH as a coenzyme. About 11% of African Americans have a deficiency of glucose 6-phosphate dehydrogenase that can lead to hemolytic anemia. (a) Using appropriate structures, write the reaction equation for the regeneration of glutathione from GSSG. (b) Glucose 6-phosphate dehydrogenase is an enzyme in what pathway discussed in this chapter? (c) Why does a deficiency of this enzyme possibly lead to hemolytic anemia?

4. Glycogen phosphorylase *a*, the active form, is allosterically inhibited by glucose and caffeine. How does this action of caffeine relate to the more familiar action of stimulation people usually experience from drinking coffee and other beverages that contain caffeine?

5. Suppose a particular glycogen molecule contained 6000 glucose residues. If a branch of eight residues occurred every eight residues of the glycogen polymer, (a) approximately how many chain ends would there be? (b) Are the ends of these chains reducing ends, nonreducing ends, or a mixture of the two?

15

Electron Transport and Oxidative Phosphorylation

I. KEY TERMS

With the help of this study guide, your textbook, and class notes, you should be able to define and explain the significance of the following terms:

oxidative phosphorylation

mitochondrial matrix

uncoupling

Q cycle

binding-change mechanism

respiratory electron transport chain

chemiosmotic theory

proton motive force

crossover point

P:O ratio

II. EXERCISES

A. True–False

_____ 1. Cytochromes are iron-containing proteins that transfer electrons in the respiratory chain.

_____ 2. Cytochrome c oxidase is the respiratory complex that directly transfers electrons to oxygen.

_____ 3. All components of the respiratory chain are proteins.

_____ 4. Each complex of the respiratory electron transport chain contains iron sulfur clusters that can perform one-electron transfers.

_____ 5. In intact mitochondria, a respiratory inhibitor, such as rotenone, stops electron transport but has no effect on ATP production.

_____ 6. The cytochrome c oxidase complex utilizes copper ions as well as iron ions to achieve the transfer of electrons to oxygen.

_____ **7.** 2,4-Dinitrophenol allows increased oxygen consumption even when mitochondria are deprived of ADP.

_____ **8.** Oxidation of one mole of succinate by the electron transport system produces three moles of ATP.

_____ **9.** The production of ATP requires energy that is supplied by a proton concentration gradient.

_____ **10.** The inner membrane knobs contain an enzyme that, in vitro, was shown to catalyze the conversion of ATP to ADP and P_i.

B. Short Answers

1. The heme-containing respiratory pigments are the _____.

2. Groups of solubilized proteins and other factors isolated from mitochondria that individually catalyze reactions from a segment of the respiratory chain are called _____.

3. The number of respiratory complexes in the electron transport chain is _____. These complexes contain multiple protein subunits and various other _____ that are directly involved in electron transfer.

4. The location in the respiratory electron-transport chain where a carrier changes from a reduced state to an oxidized state is called a/an _____.

5. The only electron-transferring component of the respiratory chain that is a lipid is _____.

6. Two inhibitors mentioned in your text that inhibit the transfer of electrons to oxygen by cytochrome _c_ oxidase are _____ and _____.

7. The respiratory complex intimately connected with the citric acid cycle is _____. This complex contains _____, a citric acid cycle enzyme.

8. The machinery for ATP synthesis is located in the _____.

9. The theory proposed by _____ suggests that ATP formation is driven by the proton concentration gradient across the inner mitochondrial membrane. This hypothesis is called the _____.

10. The ratio of the moles of ATP produced per mole of oxygen atom reduced in the electron transport chain is called the _____.

C. Problems

1. Write the net reaction for the oxidation of NADH by the electron transport-oxidative phosphorylation process.

2. Write the net reaction for the oxidation of succinate by the electron transport-oxidative phosphorylation process.

3. Why does the oxidation of succinate lead to the formation of only about 1.5 ATP per mole rather than 2.5 ATP per mole as does NADH?

4. Using information in text Table 15.2, determine the free energy change for the transfer of a pair of electrons from NADH to the FMN of Complex I.

5. Explain how each of the following is affected when antimycin A is added to functioning mitochondria:

   (a) oxygen consumption   (b) ATP production

   (c) $NAD^{\oplus}$ regeneration   (d) heat production

6. Explain the effects of 2,4-dinitrophenol on mitochondria with respect to the same four processes listed in the previous question.

7. In Section 11.9.B of your text, the standard free energy change for the transfer of electrons from NADH to oxygen was calculated to be –220 kJ/mol. Verify this value with the information given in text Table 15.3.

8. What percent of the standard free energy change is conserved as ATP by the electron transport-oxidative phosphorylation process when NADH is oxidized?

9. What is the mechanism by which 2,4-dinitrophenol causes uncoupling of oxidative phosphorylation from the electron transport process?

10. What is the role of transport proteins in mitochondrial synthesis and cytoplasmic consumption of ATP?

D. Additional Problems

1. In the 1940s, 2,4-dinitrophenol was briefly used as an agent for weight reduction. Some individuals experienced elevated temperature, a marked increase in oxygen consumption, weakness, and profuse sweating. Some fatalities led to a ban of this agent for weight reduction. (a) Explain how this agent might achieve weight reduction.   (b) Explain the symptoms observed in persons treated with 2,4-dinitrophenol.

2. Respiratory Complex IV, cytochrome c oxidase, catalyzes the transfer of electrons to oxygen. The proton concentration gradient is also affected by this transfer of electrons. "The effect is the same as a net transfer of four $H^{\oplus}$ for each pair of electrons." (a) Write a reaction equation for the reduction of one mole of molecular oxygen by the action of Complex IV. (b) Explain the above statement in quotes.

3. Show that the transfer of electrons from NADH to FMN of Complex I, under standard conditions, does not provide enough energy to form ATP but that the transfer of electrons from NADH to coenzyme Q does supply the required energy for one ATP to be formed.

4. Thermogenesis occurs in the brown adipose tissue of hibernating animals. What does the process do? What mechanism is involved and what stimulates the mechanism?

5. Imagine that you have isolated fragments of the mitochondrial inner membrane. You believe that these fragments contain one or more of the four respiratory complexes. By additional experimentation, you find that neither succinate nor NADH is oxidized, in the presence of Q, by the isolated fragments but that the following reaction is catalyzed:

$$2 \text{ Cyt } c(Fe^{2\oplus}) + {}^1/_2 O_2 + 2 H^{\oplus} \rightarrow 2 \text{ Cyt } c(Fe^{3\oplus}) + H_2O$$

What respiratory complex (or complexes) is (are) in your isolated fragments?

16

Photosynthesis

I. KEY TERMS

With the help of this study guide, your textbook, and class notes, you should be able to define and explain the significance of the following terms:

photosynthesis

light reactions

chloroplast

stroma

grana

accessory pigment

reaction center

light-harvesting complex

photophosphorylation

reductive pentose phosphate cycle

Rubisco

C_4 pathway

Crassulacean acid metabolism

phototroph

dark reactions

thylakoid membrane

lumen

chlorophyll

photosystem

special pair

Z-scheme

cyclic electron transport

stomata

photorespiration

II. EXERCISES

A. True–False

_____ **1.** Photosynthetic eukaryotes usually have both mitochondria and chloroplasts that contribute to energy production.

_____ **2.** The light reactions of photosynthesis occur only in daylight and the dark reactions of photosynthesis occur only at night.

_____ **3.** Oxygen is produced as a consequence of the dark reactions of photosynthesis.

_____ **4.** The oxygen produced by green plants comes directly from the oxygen atoms of CO_2 taken in by the plant.

_____ **5.** Although chlorophyll *a* and chlorophyll *b* have specific absorption maxima, plants are able to absorb and use energy across the visible spectrum.

_____ **6.** All photosynthetic organisms rely on either chlorophyll *a* or chlorophyll *b* to absorb light.

_____ **7.** A protonmotive force is created in chloroplasts as protons are pumped across the thylakoid membrane into the stroma.

_____ **8.** The functional units of photosynthesis in plants are the photosystems, which are complexes of pigment molecules and proteins located in the thylakoid membrane.

_____ **9.** The light reactions produce one ATP and one NADPH for each pair of electrons transferred from water, but, in the dark reactions, two molecules of NADPH and three molecules of ATP are consumed in the reduction of one molecule of CO_2 to carbohydrate.

_____ **10.** In plants, sucrose and starch are both formed in the cytosol.

B. Short Answers

1. The two major processes of photosynthesis are the _____ reactions, which produce _____, _____, and _____, and the _____ reactions, which use CO_2, ATP, and NADPH to produce_____.

2. In eukaryotic organisms, photosynthesis occurs in the organelles called _____.

3. Within a chloroplast, the _____ membrane is arranged in stacks called _____.

4. Protein-bound pigment molecules that harvest light for the photosystems in the thylakoid membrane are called _____.

5. The process by which ATP is formed as photosynthesis occurs is called _____.

6. Each photosystem contains a _____, which is a complex of proteins, electron-transporting cofactors, and two chlorophyll molecules called the _____.

7. In a chloroplast, the light-dependent reactions take place in the _____.

8. An iron-sulfur protein coenzyme involved in the transfer of electrons that achieves the reduction of $NADP^{\oplus}$ to NADPH is _____.

9. The _____ is associated with photosystem II on the lumen side of the granal lamella and is directly involved with the production of oxygen from water.

10. The compound formed when CO_2 is fixed into an organic product in the RPP cycle is _____.

11. The cofactor required for the reduction of 1,3-*bis*phosphoglycerate to glyceraldehyde 3-phosphate is _____.

12. The enzyme that catalyzes the formation of 3-phosphoglycerate from CO_2 and ribulose 1,5-*bis*phosphate is called _____, also known as _____.

C. Problems

1. Summarize the events that occur in each of the two major processes of photosynthesis.

2.. Calculate the number of moles of water that must be oxidized in photosynthesis to provide enough reducing power, as NADPH, to form one mole of sucrose.

3. Prove that light of shorter wavelengths provides a greater amount of energy for photosynthesis than does light of longer wavelengths. Compare the energy of a mole of photons of light of 700 nm to that of light of 400 nm. Recall that the energy of a photon is $E = h\nu$, where h is Planck's constant (6.626×10^{-34} J s), and ν is the frequency of the light, related to the speed of light and its wavelength, λ, by $c = \lambda\nu$. The speed of light, c, is 3.00×10^7 nm/s.

4. Explain the role of carotenoids in photosynthesis.

5. Indicate at least three ways in which photosynthesis is similar to electron transport and oxidative phosphorylation.

6. Emerson and Arnold found that oxygen yields of *Chlorella* (a green algae) subjected to flashes of light leveled off at light intensities that were sufficient to excite only a small fraction of 2400 chlorophyll molecules. Additionally, they found that 8 photons of energy must be sequentially absorbed to produce one oxygen molecule. They described a functional cluster as the number of chlorophyll molecules that utilize one photon of energy. (a) How many chlorophyll molecules comprise a functional cluster? (b) How many functional clusters are needed to produce one oxygen molecule? (Note: This work gave rise to the concept of a photosynthetic unit or a photosystem.)

7. The amount of ATP produced in the light-dependent reactions of photosynthesis does not seem adequate to fill the ATP requirements of the dark (light-independent) reactions. How are the levels of NADPH and ATP balanced by the light reactions at the relative levels of their use in CO_2 fixation?

8. What is different about the locations of photosystem I, PSI, and photosystem II, PSII? What is the functional importance of this difference?

9. In 1965, Jagendorf and Uribe studied ATP formation in spinach chloroplasts. In one study, they soaked the chloroplasts in pH 4.0 buffer, which allowed the pH to drop to 4.0 inside the chloroplasts, then quickly changed the external pH to 8.0. This resulted in the formation of ATP from ADP and P_i. How did this study contribute to our present understanding of photosynthesis?

10. (a) How many molecules of CO_2 are required for photosynthetic hexose formation? (b) Explain how the CO_2 molecules are utilized.

D. Additional Problems

1. In a photosystem, both the antenna complex and the reaction center contain chlorophyll molecules. Only the special pair chlorophyll molecules in the reaction center are intimately involved in the conversion of light energy to chemical energy. Speculate on the difference in the chlorophylls in the reaction center from those of the antenna complex that allows them to perform different functions. (Note: You may need to consult sources other than your text to answer this question.)

2. Figure 16.1 shows a plot of the 1966 data of Jagendorf and Uribe in which the ATP yield of spinach chloroplasts buffered at pH 4.0 was determined when their external (second stage) pH was increased. The chloroplasts had previously been illuminated while buffered at pH 4.0. How does this yield of ATP per chlorophyll compare with the findings of Emerson and Arnold discussed earlier (Problem C.6.)?

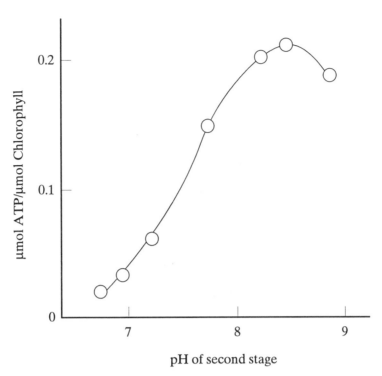

Figure 16.1 ATP Production of Spinach Chloroplasts Subjected to Different pH Gradients.

3. The enzyme Rubisco accepts both CO_2 (K_m = 20 μM, k_{cat} = 3 s^{-1}) and O_2 (K_m = 200 μM) as substrates, which compete for the same active site. (a) What are the normal relative rates of the two processes in vivo? (b) Are there circumstances under which the competing reactions can limit the growth of the plant? (c) Photorespiration in plants with the C4 pathway (corn, sorghum, sugarcane) is minimal and does not limit growth of these plants. By what process is photorespiration minimized in these plants? (Note: You may have to consult sources other than your text to answer all parts of this question.)

4. If solar energy of about 7 J $(cm^2 s)^{-1}$ is absorbed by a 10–cm^2 area of a leaf and results in the production of 5.53 grams of glucose in one hour, what is the percent efficiency of the hexose production process? For the in vitro oxidation of glucose, $\Delta G^{0'}$ = –686 kcal/mol or 2870 kJ/mol.

5. The following equation has been used to calculate the free energy change for the transfer of protons across the inner mitochondrial membrane to the intermembrane space.

$$\Delta G = 2.3 \text{ RT } [\text{pH(in)}-\text{pH(out)}] + ZF\Delta\Psi,$$

where R is the gas constant, T is the absolute temperature, Z is the charge on the proton, F is the Faraday and $\Delta\Psi$ is the membrane potential. Assume a temperature of 25°C, [ATP]/[ADP][P_i] = 1000, and determine the pH difference across the thylakoid membrane necessary to drive the formation of ATP during photosynthesis. Assume that a transfer of 3 protons is involved and that the membrane is freely permeable to $Ca^{2\oplus}$ and $Cl^{\ominus}$ ions.

6. In 1945, Melvin Calvin and coworkers showed that 3-phosphoglycerate was the first compound to show a ^{14}C-label when $^{14}CO_2$ was used in their studies of photosynthesis. (a) What did this result suggest about the size of the carbon chain of the CO_2 acceptor? (b) How was this conclusion modified when it was discovered that there were two moles of 3-phosphoglycerate formed for each mole of CO_2 fixed? (c) Where is the label located in 3-phosphoglycerate?

17

Lipid Metabolism

I. KEY TERMS

With the help of this study guide, your textbook, and class notes, you should be able to define and explain the significance of the following terms:

triacylglycerol

β-oxidation pathway

ketogenesis

citrate transport system

local regulator

acidic phospholipid

lipoprotein

carnitine

acyl carrier protein

eicosanoid

neutral phospholipid

isopentenyl pyrophosphate

II. EXERCISES

A. True–False

_____ **1.** L-Carnitine is an intermediate in the β-oxidation pathway.

_____ **2.** Most naturally occurring fatty acids have even-numbered carbon chains.

_____ **3.** Fatty acids are degraded two carbons at a time from the carboxyl end of the molecule.

_____ **4.** β-Oxidation of eicosanoyl CoA yields ten NADH, ten QH_2, and ten acetyl CoA.

_____ **5.** The activation of fatty acids required for β-oxidation is an energy-consuming process.

_____ **6.** The mitochondrial matrix is the site of both the degradation and synthesis of fatty acids.

_____ **7.** The liver is the primary site of ketone body formation.

_____ **8.** Almost all of the NADPH required for the synthesis of fatty acids is furnished by the pentose phosphate pathway.

_____ **9.** The carbon atoms of the steroid ring structure of cholesterol are all derived from acetyl CoA.

_____ **10.** A high ratio of LDL to HDL in the blood is associated with increased risks of cardiovascular disease.

_____ **11.** Animal cells have the ability to create double bonds in synthesized fatty acid chains but still require polyunsaturated fatty acids from dietary sources.

B. Short Answers

1. The products of β-oxidation of fatty acids are _____, _____, and _____.

2. In β-oxidation, the enzyme that catalyzes cleavage of its substrate to produce acetyl CoA as a product is _____.

3. The molecules known as ketone bodies are _____, _____, and _____.

4. The eukaryotic organelles in which β-oxidation occurs are _____, and _____.

5. Acetyl CoA is required for fatty acid synthesis, but the form through which carbons are supplied for extending fatty acid carbon chains is _____.

6. Enzymes that create double bonds in saturated fatty acyl CoA molecules are called _____.

7. Short-lived regulatory molecules, called _____, include prostacyclin, thromboxane A_2, and other prostaglandins that are derived from prostaglandin H_2, an eicosanoid derived from _____.

8. In addition to the prostaglandins and thromboxanes formed as one class of eicosanoids, another class includes the _____, molecules involved in allergic response.

9. _____ is the name of a drug used to lower blood cholesterol levels. It is a competitive inhibitor of HMG-CoA reductase, the enzyme that catalyzes the first committed step in cholesterol biosynthesis.

10. Non-cyclic precursors of cholesterol that have 10, 15, and 30-carbon chains, respectively, are _____, _____, and _____.

C. Problems

1. Write the overall (net) equation for the β-oxidation of stearoyl CoA.

2. The activation of a fatty acid molecule is said to consume two ATP equivalents. Why is that, since only one ATP molecule is used?

3. Why are the natural unsaturated fatty acids not suitable substrates for the enzymes of the β-oxidation pathway, when one intermediate of the pathway is unsaturated?

4. Under what conditions do liver mitochondria synthesize acetoacetate and β-hydroxybutyrate from some of the acetyl CoA formed from fatty acids?

5. If ^{14}C-labeled bicarbonate is used in cellular fatty acid biosynthesis, where does the label appear after a fatty acid like palmitate has been synthesized?

6. Cholestyramine is a non-digestible anion-exchange resin that has been taken orally to help lower human cholesterol levels. How does it cause blood cholesterol levels to decrease?

7. Your text shows that, compared to the ATP yield of glucose, palmitate yields about 24% more ATP, on a per carbon basis. Is this amount of extra ATP yield the same for all saturated fatty acids? To determine the answer, calculate the ATP yield for the oxidation of stearic acid and compare it to that of glucose, on a per carbon basis.

8. (a) What are the functions of the citrate transport system? (b) Is energy required for or produced by this system?

9. Write two separate reaction equations, one for each step in which energy is used or produced in the citrate transport system.

10. What is the key regulatory enzyme in fatty acid synthesis and what reaction does it catalyze?

11. You have two containers whose labels are missing. You know that one solution contains cholesterol and the other contains sphingomyelin. Suggest a simple laboratory procedure that would allow you to identify the solution that contained cholesterol.

D. Additional Problems

1. Is there a biochemical basis for the statement that going on a carbohydrate-free diet can lead to a life-threatening situation?

2. Release of fatty acids for energy production is inhibited by insulin. In diabetics that lack insulin, massive amounts of fatty acids are released from adipose tissue. This and other factors lead to an acceleration of fatty acid oxidation. Can fatty acid oxidation proceed indefinitely? What are the consequences?

3. Why are lovastatin and cholestyramine ineffective in lowering cholesterol levels in individuals with familial hyper-cholesterolemia (FH)? (In individuals with FH, LDL-receptor proteins are produced at lower than normal levels, or not at all in those individuals severely afflicted.)

4. Early studies showed that the protein avidin strongly inhibited the incorporation of acetyl CoA into fatty acid. What is the nature of this interference?

5. As a general rule, organisms without a glyoxylate cycle cannot convert saturated fatty acids to carbohydrates. What is a minor exception to this rule and why is it a minor exception?

18

Amino Acid Metabolism

I. KEY TERMS

With the help of this study guide, your textbook, and class notes, you should be able to define and explain the significance of the following terms:

essential amino acid	glucose-alanine cycle	cumulative feedback inhibition
nitrogen fixation	ketogenic	urea cycle
turnover	nonessential amino acid	glucogenic

II. EXERCISES

A. True–False

_____ 1. In mammals, only about one fourth of the amino acids used for protein synthesis are essential (must be obtained in the diet).

_____ 2. Amino acids derived from ingested proteins in the diet of mammals are used exclusively for cellular protein synthesis.

_____ 3. The majority of transaminases exhibit a specificity for L-glutamate and α-ketoglutarate.

_____ 4. Urea is synthesized in mammalian kidneys.

_____ 5. Nitrogen fixation, catalyzed by the enzyme nitrogenase, is a process that requires energy.

_____ 6. Glutamine is a nitrogen donor in several biosynthetic reactions and represents a means of carrying nitrogen between tissues.

_____ **7.** Each of the 20 common amino acids can be described as being either glucogenic or ketogenic.

_____ **8.** Arginase is an enzyme found at high activity levels in nearly all mammalian organs and tissues.

_____ **9.** Protons from acids generated during metabolism are neutralized by bicarbonate in the blood. Bicarbonate must be regenerated, however, so that blood pH is maintained by the bicarbonate buffer system.

_____ **10.** Nitroglycerin exerts its effect, dilation of coronary arteries, by virtue of being metabolized to nitric oxide.

B. Short Answers

1. A coenzyme required by the transaminases is _____. Its function in transamination is _____.

2. While most of the common amino acids can be formed by transamination, little or none of the two amino acids _____, and _____ is formed by transamination.

3. Amino acids whose carbon chains are catabolized to form acetyl CoA are said to be _____. Those that form intermediates of glycolysis or the citric acid cycle are said to be _____.

4. A mutation in the gene that codes for phenylalanine hydroxylase makes afflicted individuals unable to tolerate normal quantities of phenylalanine. The resulting disease is known as _____.

5. There are nine cellular metabolites that inhibit glutamine synthetase. No single metabolite inhibits the enzyme completely, but the degree of inhibition increases as additional inhibitors bind. This process is known as _____.

6. The sources of the two nitrogen atoms and the carbon atom of urea are _____, _____, and _____, respectively.

7. One of the nitrogen atoms and the carbon atom of urea enter the urea cycle as a single molecule, namely _____.

8. The urea cycle requires energy to form urea. The equivalent of _____ ATP molecules are required.

9. The exchange of glucose and alanine between muscle and liver is called the _____.

10. As serine is catabolized, it is converted to glycine, which is broken down by the action of _____.

C. Problems

1. What are the two major stages by which amino acids are degraded?

2. In the glucose-alanine cycle, alanine formed in muscle is transported by the blood to liver where it is acted on by alanine transaminase. Using appropriate structures, write this transamination reaction.

3. To regenerate α-ketoglutarate required by transaminases, L-glutamate must be oxidatively deaminated in a reaction catalyzed by glutamate dehydrogenase. How can this deamination occur when the $\Delta G^{0\prime}$ for the reaction is +27 kJ/mol?

4. The answer to problem II.B.8. indicated that the equivalent of 4 ATP molecules is required for the synthesis of one molecule of urea. Detail this energy expenditure.

5. Discuss the regulation of the urea cycle.

6. Explain/criticize the statement that glycine synthesis is dependent upon levels of folate and serine.

7. In text Section 18.4.C., you are told that proline is formed as a result of nonenzymatic cyclization of glutamate γ-semialdehyde to form a Schiff base that is then reduced. Using appropriate structures, write the reaction sequence for the formation of proline from glutamate γ-semialdehyde. You may use the appropriate number of electrons (as H atoms) rather than some specific reducing agent.

8. List seven compounds of central metabolism into which carbon atoms from the carbon chains of amino acids are channeled.

9. Separately list those amino acids that are ketogenic, those that are glucogenic and those that fall into both categories.

10. Explain how the synthesis of two nonessential amino acids is dependent upon two essential amino acids.

D. Additional Problems

1. Imagine that you worked with Sir Hans Krebs during the time he was elucidating the urea cycle. You used slices of rat liver tissue and exposed these tissues to a solution containing urease. The gas evolved was collected and found to be 2.4 mL at STP conditions. How many moles of urea were present in the tissue slices? (Assume that none of the gas produced dissolved in the solution.)

2. Compare the function of the alanine-glucose cycle to that of the Cori cycle. Under what circumstances might each operate?

3. The ammonium ion is unable to enter the brain, but several known defects in urea cycle enzymes are known to lead to mental retardation. What is the connection?

4. The first nitrogen atom of the urea molecule enters the urea cycle as carbamoyl phosphate, formed from bicarbonate and ammonium ions. Detail the entry of the second nitrogen atom of the urea molecule and how the nitrogen carrier itself obtained the nitrogen.

5. When ^{18}O-labeled citrulline (amide oxygen) reacts with aspartate to form argininosuccinate, as catalyzed by argininosuccinate synthetase, (a) where does the label appear? (b) What does this suggest about a possible intermediate of the reaction? (Note: Sources other than your text may be required to answer this question.)

19

Nucleotide Metabolism

I. KEY TERMS

With the help of this study guide, your textbook, and class notes, you should be able to define and explain the significance of the following terms:

5-phosphoribosyl 1-pyrophosphate   inosine 5′-monophosphate   purine nucleotide cycle

II. EXERCISES

A. True–False

_____ 1. The nitrogenous bases of nucleotides and nucleosides are of two types—the purines, which are heterocycles containing six atoms, and the pyrimidines, which are heterobicycles containing nine atoms.

_____ 2. Nucleotide biosynthesis occurs in virtually all tissues.

_____ 3. To synthesize the purine ring, liver cells require metabolic sources of glycine, formate, glutamine, CO_2, and aspartate.

_____ 4. Tissues synthesize the purine ring and then attach it to ribose 5-phosphate to make nucleotides.

_____ 5. Inosine 5′-monophosphate (IMP) is an intermediate from which both 5′-AMP and 5′-GMP are synthesized.

_____ 6. GTP is required for the synthesis of AMP (and subsequently ATP), and ATP is required for the synthesis of GMP (and subsequently GTP).

_____ 7. To synthesize the pyrimidine ring, liver cells require aspartate, glutamine and bicarbonate.

_____ **8.** Orotate is the end product of the pyrimidine biosynthetic pathway and is the derivative from which the pyrimidine nucleosides and nucleotides are made.

_____ **9.** Like the purine ring system, the pyrimidine ring system is formed using ribose 5-phosphate as the foundation upon which it is built.

_____ **10.** dTMP is formed by the reduction of the ribose moiety of ribothymidylate.

_____ **11.** Most free purine and pyrimidine molecules are salvaged, but some are catabolized. Catabolism of both purine and pyrimidine molecules leads to the excretory product, uric acid.

B. Short Answers

1. The active form of ribose 5-phosphate required for nucleotide biosynthesis is _____.

2. In liver cells, the location of purine biosynthesis is _____.

3. Of the two fused rings of purine nucleotides, the one formed first in the cell is _____.

4. Four feedback inhibitors of enzymes of the purine biosynthetic pathway are _____, _____, _____, and _____.

5. Carbamoyl phosphate is involved in the biosynthesis of which type of nucleotides, the pyrimidines or the purines? _____

6. _____ is the product of purine nucleotide catabolism that is involved in gout.

7. CTP is formed from UTP by transformation of the keto group at C-4 of the pyrimidine ring to an amino group. The source of the amino group is _____.

8. Two enzymes, sometimes called salvage enzymes, that are responsible for recycling the bases adenine, hypoxanthine, and guanine to the corresponding nucleotides AMP, IMP, and GMP are _____ and _____.

9. A pathway in muscle that produces ammonia and fumarate, under exercising/work conditions, and involves nucleotides is called the _____.

10. Two thioesters of central metabolism formed from the catabolism of pyrimidines are _____ and _____.

C. Problems

1. List the products, and the number of moles of each, formed by the hydrolysis of (a) cytidine, (b) inosine monophosphate, and (c) dGTP.

2. Glutamine is required in the biosynthetic pathways of both the purines and the pyrimidines. What is its involvement in purine biosynthesis?

3. In general, biosynthesis is energy-consuming. Starting with ribose 5-phosphate, how many ATP equivalents are consumed in the synthesis of 5′-AMP? Assume sufficient glutamine and other required materials are available. Also ignore any reduced cofactors formed that could provide ATP upon reoxidation.

4. Glutamine-PRPP amidotransferase catalyzes the first committed step of purine nucleotide synthesis. It is inhibited by both guanosine and adenosine phosphates (feedback inhibition). Discuss the likelihood of: (a) more than one inhibitor-binding site on the enzyme, (b) complete inhibition by any one of the inhibitors alone.

5. Write the overall (net) equation for the cellular formation of UMP. (Structures are not required.)

6. Step 6 in the biosynthesis of IMP involves a carboxylation catalyzed by aminoimidazole ribonucleotide carboxylase (AIR carboxylase). What is unusual about the requirements of this enzyme?

7. Using appropriate structures, write the equation for the formation of CTP from UTP.

8. If 3-[^{14}C]-serine is supplied to cells performing nucleotide biosynthesis, where will the label appear in dTMP?

9. Explain how deoxyribonucleotides are synthesized from ribonucleotides.

10. (a) What hereditary deficiency leads to Lesch-Nyhan syndrome? (b) Why is the disease usually restricted to males? (c) What are the clinical symptoms of the disease? (d) Why do afflicted individuals have greatly increased *de novo* synthesis of purine nucleotides?

D. Additional Problems

1. Assume that ATP, glutamine, and ^{14}C-$HCO_3^{\ominus}$ are incubated with tissue slices that contain the enzymes and other materials necessary for pyrimidine nucleotide biosynthesis. If orotidine 5′-phosphate (OMP), UMP, and CO_2 are isolated from the system, in which of these should the label be found?

2. The purine nucleotide cycle provides the citric acid cycle intermediate fumarate to assist operation of the citric acid cycle and production of ATP for energy needs. $NH_4^{\oplus}$ is also produced by the purine nucleotide cycle. If all $NH_4^{\oplus}$ so produced is eliminated as urea, what is the energy cost per mole of fumarate produced by the purine nucleotide cycle?

3. What would be the consequences of a hereditary deficiency of the enzyme deoxyuridine triphosphate diphosphohydrolase (dUTPase)?

4. von Gierke's glycogen storage disease is caused by a deficiency of glucose 6-phosphatase. This can lead to overproduction of uric acid and to gout. Explain the connections.

5. Azaserine is a glutamine analog that has been used as an affinity label. (a) What nucleotide biosynthetic enzymes would be affected by administration of azaserine or similar antibiotics? (b) What is the mechanism of such inhibition?

$$
\begin{array}{c}
COO^{\ominus} \\
| \\
H_3N^{\oplus}\!-\!C\!-\!H \\
| \\
CH_2 \\
| \\
O \\
| \\
\underset{O}{\overset{}{\diagup}}C \\
\diagdown CH = N^{\oplus} = N^{\ominus}
\end{array}
$$

20

Nucleic Acids

I. KEY TERMS

With the help of this study guide, your textbook, and class notes, you should be able to define and explain the significance of the following terms:

nucleic acid
genome
translation
base pair
antiparallel
major groove
rise
kilobase pair
melting point
A-DNA
supercoiling
ribosomal RNA
transfer RNA
small RNA
chromatin
chromosome

nucleosome
solenoid
ribonuclease
exonuclease
restriction endonuclease
palindrome
hemimethylated
polynucleotide
transcription
B-DNA
double helix
stacking interactions
minor groove
pitch
denaturation

melting curve
Z-DNA
topoisomerase
ribosome
messenger RNA
hairpin
30-nm fiber
histone
core particle
nuclease
deoxyribonuclease
endonuclease
restriction methylase
sticky end
restriction map

II. EXERCISES

A. True–False

_____ **1.** Nuclein was the term used by Miescher for the mixture of nucleic acids and protein isolated from white blood cells.

_____ **2.** Chargaff determined that the molar ratios of adenine to thymine and of guanine to cytosine in DNA are about one. This means that the sum of the adenine + thymine bases will be equal to the sum of the guanine + cytosine bases.

_____ **3.** The Watson-Crick model of DNA is a right-handed helix of two separate chains, both oriented in the same direction.

_____ **4.** The ultraviolet absorbance of one mole of duplex (double stranded) DNA is greater than that of the same amount of denatured DNA.

_____ **5.** Endonucleases are enzymes that catalyze the cleavage of phosphodiester linkages in the interior of nucleic acids.

_____ **6.** Some eukaryotic chromosomes are circular.

_____ **7.** DNA can be hydrolyzed in the laboratory by dilute NaOH.

_____ **8.** Hydrolysis of RNA by the catalytic action of pancreatic ribonuclease or by dilute NaOH involves formation of a 2′,3′-cyclic nucleoside monophosphate intermediate.

_____ **9.** In many bacteria, there are specific restriction methylases and restriction endonucleases that recognize the same sequence of bases in the substrate DNA.

_____ **10.** Sticky ends are pieces of single-stranded DNA formed by the action of certain endonucleases as a result of duplex DNA denaturation.

B. Short Answers

1. The hydrated, more common form of DNA is _____.

2. The type of DNA that exists as a left-handed helix is _____.

3. The allowed base pairing in DNA is thymine with _____ and cytosine with _____.

4. The process of using heat to dissociate duplex DNA into separate single strands is called _____. The temperature at which half of the DNA has become single stranded is called the _____.

5. Bacterial enzymes that recognize and catalyze the cleavage of foreign DNA are called _____.

6. The DNA sequences recognized by most restriction endonucleases are called _____.

7. Eukaryotic DNA has a number of basic proteins called _____ associated with it. The resulting nucleoprotein complex is called a/an _____.

8. Four types of interactions that influence the stability of duplex DNA are _____, _____, _____, and _____.

9. Enzymes that break DNA, unwind or overwind the double helix, and rejoin the strands are called _____.

10. RNA molecules that carry information from DNA to the ribosomes for protein synthesis are called _____.

C. Problems

1. The guanine content of a double-stranded DNA sample was determined to be 17% (of total bases in the sample). Name the other bases present in the DNA and the corresponding percentages of each.

2. In a B-DNA segment that is about 51 nm in length, what is the total number of bases?

3. Alexander Rich synthesized hexadeoxynucleotide chains with repeating cytosine and guanine bases. The association of two such strands forms a double helix with the Z-DNA conformation. These two strands are said to be self-complementary. Illustrate this self-complementary aspect.

4. Of the total bases present, what is the approximate mole% of adenine in a DNA sample with a T_m of 90°C, based on the data presented in Figure 20.1.

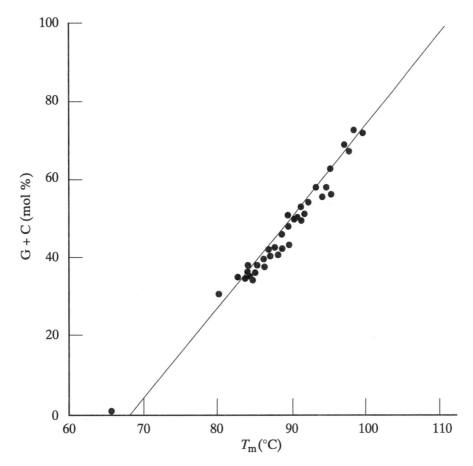

Figure 20.1 Melting Temperature Variation with DNA Base Content

5. What is the expected T_m for a DNA sample with a thymine content of 20%?

6. Explain the mechanism by which some bacteria recognize and destroy foreign DNA without destroying their own DNA.

7. Why does duplex DNA form a helix rather than existing as two side-by-side antiparallel strands?

8. What is the most abundant type of RNA in cells and what is its function?

9. What is the composition of a nucleosome and what is its relationship to the 30-nm fiber?

10. Consider the restriction endonucleases in text Table 20.3. Which enzymes do not produce sticky ends when cleaving DNA.

D. Additional Problems

1. If the largest human chromosome contains 2.4×10^8 base pairs, (a) how many nucleosomes does it contain? (b) If this chromosome contained 2000 loops of the 30-nm fiber as it is attached to an RNA-protein scaffold, how many nucleosomes are in each loop, on the average?

2. You are a biochemist studying DNA isolated from a particular type of phage. You note that the DNA serves as a substrate for DNase I and DNase II but not for snake venom phosphodiesterase or spleen phosphodiesterase. Further, analysis of base content indicates that the ratio of thymine to adenine is 1.36. Explain these data. (Table 20.1 contains information that may be helpful in the solution of this problem.)

Table 20·1 Some representative nucleases

	Substrate	Product
Exonucleases (3′→5′)		
Exonuclease I (*E. coli*)	Single-stranded DNA	Nucleoside 5′-phosphates
DNA polymerase I (*E. coli*)	Single-stranded DNA*	Nucleoside 5′-phosphates
Exonuclease III (*E. coli*)	Double-stranded DNA*	Nucleoside 5′-phosphates
Bacillus subtilis exonuclease	Double-stranded DNA, RNA	Nucleoside 5′-phosphates, oligonucleotides
Snake venom phosphodiesterase	Single-stranded DNA or RNA	Nucleoside 5′-phosphates (no base specificity)
Exonucleases (5′→3′)		
Bacillus subtilis exonuclease	Single-stranded DNA	Nucleoside 3′-phosphates (also works from 3′ end)
Exonuclease VII (*E. coli*)	Single-stranded DNA	Oligonucleotides (also works from 3′ end)
Neurospora crassa exonuclease	Single-stranded DNA RNA	Nucleoside 5′-phosphates
Exonuclease VI or DNA polymerase I (*E. coli*)	Double-stranded DNA	Nucleoside 5′-phosphates
Bovine spleen phosphodiesterase	Single-stranded DNA or RNA	Nucleoside 3′-phosphates (no base specificity)
Endonucleases		
Neurospora crassa endonuclease	Single-stranded DNA RNA	Polynucleosides with 5′-phosphate termini
Endonuclease I (*E. coli*)	Single- or double-stranded DNA	Polynucleosides with 5′-phosphate termini
DNase I (bovine pancreas)	Single- or double-stranded DNA	Polynucleosides with 5′-phosphate termini
DNase II (calf thymus)	Single- or double-stranded DNA	Polynucleosides with 3′-phosphate termini
RNase A (bovine pancreas)	RNA	Polynucleosides with 3′-phosphate termini

*Also catalyzes hydrolysis of nicked and gapped double-strand DNA.

3. In 1963, Jerome Vinograd studied circular DNA from polyoma virus. The DNA had the expected 1:1 ratio for A-T and G-C pairs, but two DNA bands were observed when the DNA was centrifuged. Since both types of DNA were circular and of identical base composition, why were there two bands?

4. Pancreatic DNase I catalyzes the internal cleavage of single-stranded or duplex B-DNA somewhat at random and does not cleave Z-DNA at all. Use this information to draw some conclusions about the manner in which DNase I functions. (Note that lysine and arginine residues in the enzyme are involved in binding.)

5. Nitrites in food have been a concern in recent years because they lead to the formation of nitrosamines, which appear to be carcinogenic. Additionally, nitrites can be converted to nitrous acid by stomach HCl. Nitrous acid can oxidize amino groups to keto groups on the bases of DNA and RNA. Write reactions showing the effects of nitrous acid on cytosine and adenine in DNA. Would these changes in DNA damage the cell? Explain.

21

DNA Replication, Repair, and Recombination

I. KEY TERMS

With the help of this study guide, your textbook, and class notes, you should be able to define and explain the significance of the following terms:

replication

genetic recombination

replication fork

processive

leading strand

Okazaki fragment

primosome

helicase

photoreactivation

homologous recombination

semiconservative

replisome

DNA polymerase

distributive

lagging strand

RNA primer

nick translation

direct repair

general excision-repair pathway

Holliday junction

II. EXERCISES

A. True–False

_____ **1.** The conservative model of DNA replication is supported by experimental evidence.

_____ **2.** In *E. coli,* replication is unidirectional.

_____ **3.** Bacterial replication requires dAMP, dTMP, dGMP, and dCMP.

_____ **4.** Both eukaryotic and prokaryotic cells have several different DNA polymerases that are involved in replication and repair of DNA.

_____ **5.** Kornberg's enzyme (DNA polymerase I) catalyzes the synthesis of DNA if provided with only the four deoxyribonucleoside triphosphates.

_____ **6.** DNA polymerase III performs the major role in prokaryotic replication.

_____ **7.** Eukaryotic Okazaki fragments are smaller than prokaryotic Okazaki fragments.

_____ **8.** DNA ligase plays a role in normal DNA replication as well as in repair of damaged DNA.

_____ **9.** The correction of thymine dimer formation in DNA strands requires excision of the damaged region and its replacement by a new DNA segment of the same base sequence as that removed.

_____ **10.** Requirements of normal recombination in *E. coli* are that a region of double-stranded DNA be unwound to create a single-stranded section and a segment of duplex DNA be available with a sequence the same as or similar to that of the single-stranded section.

B. Short Answers

1. The location at which duplex DNA is unwound during replication and at which new DNA is being synthesized is called the _____.

2. The DNA strand that serves as a complementary pattern along which a new DNA strand is synthesized is called the _____ strand.

3. Short segments of RNA required for synthesis of DNA on a template are called _____.

4. The DNA strand formed as a continuous strand during replication is called the _____, and the strand formed in a discontinuous process is called the _____.

5. Enzymes that remain bound to their polymeric chains (eg., DNA) during the process of many polymerization steps are said to be _____.

6. Short segments of lagging-strand DNA formed by the discontinuous process are called _____.

7. _____ is the name of the process by which DNA polymerase I catalyzes the formation of DNA segments to replace the RNA primer segments required by DNA polymerase III during replication.

8. In the process of general recombination, the structure formed as a result of exchange of single strands (strand invasion) is called the _____.

9. The protein complex in *E. coli* responsible for DNA replication is called the _____.

10. Since DNA polymerase III requires single-stranded DNA as a template, duplex DNA must be unwound by proteins called _____.

C. Problems

1. Bacterial circular DNA is said to have a contour length (end-to-end length of the stretched out native DNA molecule) of 1600 µm, but the length of the cell itself is only about 2 µm. How is this possible?

2. How does acid cause precipitation of DNA?

3. What is the proofreading function of DNA polymerase III and how is it manifested?

4. How does the bacterial replisome synthesize both strands of DNA simultaneously if the strands are synthesized in a 5′→3′ direction and the template strands of the parent DNA are antiparallel?

5. If RNA primers are required for DNA synthesis, why do DNA samples isolated after replication not contain pieces of RNA?

6. What are 2′,3′-dideoxyribonucleoside triphosphates and for what are they used?

7. Describe the general steps, similar in all organisms, by which damaged DNA is corrected by the excision-repair pathway.

8. Your text indicates that the synthesis of histones and of DNA occurs in different parts of the cell, but that they occur simultaneously, and the amount of histone is equivalent to the amount of new DNA. Explain why this is necessary for proper cell function.

9. Given the overall error rate for polymerization and the chromosome size of *E. coli* (4.6×10^6 nucleotide pairs), what is the average number of incorrect bases expected to be incorporated in a single replication of the *E. coli* chromosome?

10. (a) What is different about the mechanism of action of *E. coli* DNA ligase from that of the DNA ligase in eukaryotic cells? (b) What kinetic mechanism does the enzyme employ? (c) What is the specificity of the DNA ligases with respect to DNA base sequences recognized?

D. Additional Problems

1. Indicate how prokaryotic Okazaki fragments could be distinguished from eukaryotic Okazaki fragments.

2. From the size of the *E. coli* chromosome (text Section 21.2.C) and the size of the Okazaki fragments given in text Section 21.3.A., (a) calculate the number of Okazaki fragments (OF) formed during replication of one third of the *E. coli* chromosome. (b) Calculate the time in minutes required for formation of the Okazaki fragments.

3. *E. coli* cells were provided with a short pulse of ^{3}H-deoxythymidine. Shortly thereafter, the cells were lysed, and the DNA isolated and denatured. Gel electrophoresis of the denatured DNA indicated the presence of both long and short ^{3}H-labeled DNA fragments. (a) Is the ^{3}H-labeled DNA newly-synthesized DNA or is it DNA that already existed in the cell? (b) Explain why the ^{3}H-labeled DNA fragments were not all of the same size.

4. Figure 21.1 shows the gel electrophoretic pattern for a short piece of DNA that is to be sequenced. The Sanger method, which employs the 2′,3′-dideoxyribonucleoside triphosphates (ddNTPs), was used to obtain the gel. (See text Section 21.6.) The specific ddNTP used in each of the four trials is identified above each lane of the gel. The short primer used was 5′-TA-3′. Determine the sequence of both DNA strands of this DNA fragment being sequenced.

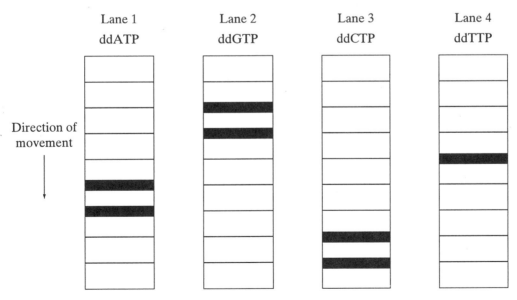

Figure 21.1 Sanger's DNA Sequencing Method.

5. Using Figure 21.1 again, give the sequence, including the primer, of each fragment in each lane of the gel.

22

Transcription and RNA Processing

I. KEY TERMS

With the help of this study guide, your textbook, and class notes, you should be able to define and explain the significance of the following terms:

gene	escape synthesis	promoter clearance
RNA polymerase holoenzyme	RNA processing	transcription bubble
ribosomal RNA	cap	pause site
transcription	intron	mRNA precursor
sigma factor	splice site	repressor
coding strand	spliceosome	inducer
consensus sequence	small nuclear ribonucleoprotein	operator
−35 region	housekeeping gene	catabolite repression
open complex	transfer RNA	splicing
termination sequence	messenger RNA	poly A tail
general transcription factor	promoter	exon
constitutive expression	operon	branch site
activator	template strand	small nuclear RNA
corepressor	TATA box	

II. EXERCISES

A. True–False

_____ 1. The unusual bases, such as N^6-methyladenylate, inosinate, and dihydrouridylate, found in tRNA molecules are incorporated during transcription of the gene.

_____ **2.** Promoter regions are found exclusively on the DNA template strand.

_____ **3.** Not all genes that encode enzymes are inducible.

_____ **4.** Protein synthesis begins on prokaryotic mRNA molecules before synthesis of a mRNA molecule is complete.

_____ **5.** It is common for eukaryotic mRNA molecules to have poly A tails.

_____ **6.** Eukaryotic mRNA molecules are capped by the addition of a 7-methylguanylate group to the 3′ end of the mRNA molecule.

_____ **7.** The 18S, 5.8S, and 28S rRNA molecules that are part of human ribosomes are all present in a single precursor rRNA molecule.

_____ **8.** The σ^{70} subunit is a component of the RNA polymerase holoenzyme that recognizes some promoter sequences.

_____ **9.** All transcription in typical eukaryotic cells is catalyzed by different RNA polymerases, found exclusively in the cell nucleus.

_____ **10.** Some mature mRNA molecules are produced by splicing, which involves removal of noncoding segments from the transcript mRNA and joining of the remaining coding regions.

B. Short Answers

1. The DNA region that acts as a transcription initiation signal is called the _____.

2. The prokaryotic holoenzyme that is involved in transcription is _____.

3. During transcription, double-stranded DNA is locally unwound. The resulting structure is called a/an _____.

4. _____ is the hexameric protein that aids in certain prokaryotic termination processes.

5. The eukaryotic enzyme that catalyzes the synthesis of mRNA precursors is _____.

6. During splicing of eukaryotic mRNA molecules, internal sequences called _____ are removed from the primary RNA transcript. Sequences that are present in the primary RNA transcript and in the mature RNA molecule are called _____.

7. The RNA molecules that combine with proteins to form the spliceosome are called _____.

8. Transcription of a negatively-regulated gene is prevented by a regulatory protein called a/an _____.

9. Transcription of a positively-regulated gene requires the presence of a regulatory protein called a/an _____.

10. Ligands that bind to and inactivate repressors are called _____.

C. Problems

1. What is the significance of the TATA box (−10 region) in bacterial DNA?

2. How does the regulatory protein rho help terminate transcription?

3. What is the *E. coli lacI* gene and what is its role in lactose use in *E. coli*?

4. What is catabolite repression?

5. What is CRP and what is the function of the CRP-cAMP complex?

6. What types of RNA are made by each of the eukaryotic RNA polymerases?

7. What is the function of capping of eukaryotic mRNA molecules?

8. The *E. coli* genome contains 4.6×10^6 base pairs (bp) and contains about 3000 genes that average 1500 bp each. This suggests that all but 2.2% of the bacterial genome is transcribed. How long would it take a single RNA polymerase to transcribe the expressed portion of the genome? (Assume that each gene is transcribed once and that initiation is not rate-limiting.)

9. (a) Why is the error rate of RNA synthesis higher than that for DNA replication? (b) Are transcription errors as serious to the cell as replication errors? Why or why not?

10. What are regions of dyad symmetry in DNA and how do they play a role in transcription?

D. Additional Problems

1. Figure 22.1 shows data obtained from *E. coli* grown on a glycerol medium. The amount of mRNA molecules encoding lactose-metabolizing enzymes is plotted against time following the addition of isopropylthiogalactoside (IPTG) to the medium. Note that glucose is added approximately three minutes after IPTG addition. What two phenomena are illustrated by these data?

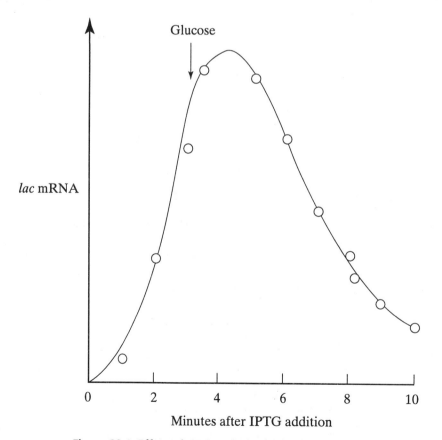

Figure 22.1 Effect of IPTG and Glucose on lac Operon.

2. It has been estimated that about 10 repressor molecules (R) per *E. coli* cell are sufficient to prevent transcription of the *lac* operon. The dissociation constant for the repressor-operator complex (RO) has been estimated to be about 10^{-13} M.

$$RO \rightleftharpoons R + O \qquad K_D = 10^{-13} \text{ M}$$

If the average cell has a volume of 0.30×10^{-12} mL and contains two copies of the *lac* operon, what is the concentration of operon (O) not bound by repressor at equilibrium?

3. Figure 22.2 shows a plot of data collected from T2-phage-infected *E. coli* in which the isolated RNA contains ^{32}P and the isolated DNA contains ^{3}H, allowing them to be separately detected. In this experiment, the DNA was denatured before being mixed with the RNA, and the components of the mixture were then separated by sedimentation in a CsCl density gradient. Note that the lower fraction numbers represent regions of higher density, removed first from the bottom of the centrifugation tube. Hall and Spiegelman (*PNAS 47*, 141 (1961)) used this type of data to support the hypothesis that DNA serves as a template for RNA synthesis. Explain how these data support the hypothesis.

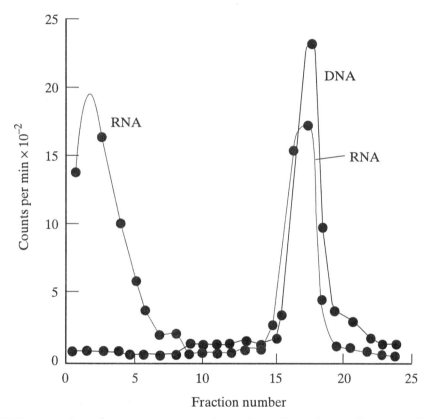

Figure 22.2 Separation of RNA and Denatured DNA by CsCl Density-Gradient Centrifugation.

4. The *E. coli* core polymerase binds DNA nonspecifically with an association constant of 10^{10} M^{-1}. Binding to the promoter region is specific when the core polymerase combines with σ^{70} to form the holoenzyme. The association constant is then 2×10^{11} M^{-1}. With this tighter binding, how is the polymerase able to leave the promoter site and transcribe the rest of the DNA template?

5. Suppose occasional errors in processing of eukaryotic mRNA transcripts occurred in which an intron was not removed. What would be the likely result at the translational level?

Protein Synthesis

I. KEY TERMS

With the help of this study guide, your textbook, and class notes, you should be able to define and explain the significance of the following terms:

codon
degenerate code
termination codon
acceptor stem
anticodon arm
D arm
wobble
isoacceptor tRNA molecule
peptidyl site
initiation factor
Shine-Delgarno sequence
polysome
elongation factor

exit site
release factor
attenuation
leader peptide
posttranslational
signal-recognition particle
protein glycosylation
reading frame
synonymous codon
initiation codon
anticodon
TΨC arm
variable arm

wobble position
aminoacyl-tRNA
aminoacyl site
initiator tRNA
polycistronic
peptidyl-tRNA
peptidyl transferase
translocation
translational regulation
leader region
cotranslational
signal peptide
signal peptidase

II. EXERCISES

A. True–False

_____ **1.** A particular aminoacyl-tRNA synthetase will recognize a specific amino acid, but will bind and amino-acylate any tRNA.

_____ **2.** Protein synthesis always begins at the amino terminus in prokaryotic systems.

_____ 3. Transfer RNA molecules are charged by the covalent attachment of the amino acid for which they are specific at the 5′ end of the tRNA.

_____ 4. N-Formylmethionyl-tRNA$_f^{Met}$ initiates protein synthesis in both prokaryotic and eukaryotic systems.

_____ 5. In bacterial systems, the 30S ribosomal subunit binds somewhere other than the actual 5′ end of the mRNA molecule to initiate protein synthesis.

_____ 6. Peptidyl transferase is a catalytic protein, found in the cell cytosol, that catalyzes peptide bond formation during translation.

_____ 7. Each new amino acid incorporated into a growing peptide chain furnishes the amino group used in the formation of a peptide bond.

_____ 8. Eukaryotic and prokaryotic mRNA molecules are polycistronic.

_____ 9. Prokaryotic mRNA molecules have Shine-Delgarno sequences for binding the small ribosomal subunit of ribosomes, but eukaryotic mRNA molecules do not.

_____ 10. Posttranslational modification of eukaryotic proteins to be secreted from the cell occurs after the proteins are secreted.

B. Short Answers

1. _____ is the term used to describe a family of tRNA molecules, each of which can be charged with the same amino acid.

2. The first codon to be determined, which coded for the amino acid _____ was _____.

3. When an aminoacyl tRNA synthetase acts to remove an amino acid mistakenly joined to a tRNA, the enzyme is exercising its _____ activity.

4. The double-stranded portion of the tRNA molecule where an amino acid is attached as the tRNA is charged is called the _____.

5. In some cases, several codons, called _____, encode the same amino acid.

6. The process by which the ribosome is shifted by one codon relative to mRNA is called _____.

7. A collection of ribosomes bound to a common mRNA template is called a/an _____.

8. The activation and incorporation of an amino acid into a growing peptide chain requires energy equivalent to the hydrolysis of _____ phosphoanhydride bonds.

9. Proteins destined for secretion are synthesized with an N-terminal sequence, called the _____, that assists in transport of the protein into the lumen of the endoplasmic reticulum, as the protein is being synthesized. This sequence is later removed.

10. Prokaryotic proteins that recognize termination codons on an mRNA molecule and cause hydrolysis of the peptidyl-tRNA are called _____.

C. Problems

1. How is an amino acid activated (made ready for incorporation into a protein)?

2. When Nirenberg and Matthaei were elucidating the genetic code, they used the supernatant fraction from a homogenized bacterial system that contained only ribosomes, tRNAs, and cytoplasmic enzymes. They added ATP, GTP, radioactive amino acids, and viral RNA. The system produced radioactive proteins except when the viral RNA was omitted. Explain.

3. Assume you had performed experiments to elucidate the genetic code. You made a synthetic RNA molecule that contained only adenine bases except for the terminal 3′ base, which was cytosine. Using this synthetic mRNA molecule along with tRNAs, amino acids, and cytoplasmic enzymes, you found that the protein produced contained all lysine residues with the exception of the C-terminal residue, which was asparagine. Assuming a triplet code, but with no code words known at that time, state three or four conclusions you could reach from the results of your experiment.

4. Is there an advantage to bacterial systems that during transcription, mRNA is synthesized from the 5′-end toward its 3′-end and that translation proceeds in a 5′→3′ direction along a template mRNA molecule?

5. What does it mean to say that the genetic code is degenerate?

6. Suppose you had been involved in determining the way in which the correct amino acids were incorporated into proteins. You wished to determine if it was the amino acid itself that was recognized by the codon of the mRNA or if it was the tRNA carrier that was recognized. To do this, you prepared cysteinyl-tRNA using ^{14}C-cysteine and cysteinyl-tRNA synthetase. You chemically transformed the radioactive cysteinyl group into an alanyl group and used the modified aminoacyl-tRNA in the synthesis of hemoglobin. After synthesis, the hemoglobin was examined for the location of radioactive residues. (a) What results would have occurred if the amino acid residue was recognized by the codon? (b) What results would have occurred if the carrier tRNA was recognized by the codon?

7. N-Formylmethionyl-tRNA$_f^{Met}$ initiates protein synthesis in many bacterial cells. (a) What prevents f-Met from being incorporated in protein chains at AUG codons at locations other than at the initiation site? (b) What prevents methionyl-tRNA^{Met} from initiating protein synthesis?

8. In prokaryotes, after the first peptide bond is formed between N-formylmethionine and the next amino acid that is brought into the A site, (a) how is the protein synthesis machinery made ready for the next (third) amino acid to be incorporated? (b) What is the name of this process?

9. Contrast translation in eukaryotic systems to that in prokaryotic systems.

10. Calculate the energy required (in kJ) for the synthesis of 0.685 mol of a protein that contains 250 amino acid residues, by the translation process, assuming standard conditions.

D. Additional Problems

1. In a particular bacterial cell, it is estimated that a single translation complex can synthesize a 300-residue polypeptide in about 20 seconds. The cell is estimated to contain about 5000 mRNA molecules at a given moment and is capable of producing about 1000 protein molecules per second. (a) For synthesis of the 300-residue protein by a single translation complex, how many mRNA nucleotides are being translated per second? (b) Based on the estimated cell production capacity, how many translation complexes, on average, are translating a particular mRNA at any given moment?

2. You are investigating the effect of various compounds on protein synthesis. You find that oligonucleotides containing Shine-Delgarno sequences inhibit prokaryotic systems but not eukaryotic systems. (a) What particular translation process was being inhibited? (b) Why are both types of systems not inhibited? (c) Would truncated proteins be formed in the prokaryotic system?

3. Assume you had synthesized an mRNA molecule that contains only uracil and guanine residues. The nucleotides were randomly incorporated and analysis showed that 76% of the bases in the mRNA were uracil and 24% were guanine. (a) What amino acid residues, and what percentages of each, would be incorporated into a protein synthesized using this molecule as a messenger? (b) If the protein contained 1000 residues, how many of the most abundant residues would there be? (c) How many of the least abundant residues would there be?

4. Binding of signal-recognition particles (SRP) to a segment of some incompletely-formed proteins prevents further translation. (a) From what is the name for the SRPs derived? (b) What happens to allow the completion of the partially synthesized protein?

5. Selenium is an essential trace element found in both prokaryotic and eukaryotic enzymes. Selenium is present in certain *E. coli* enzymes as selenocysteine residues, formed by the selenation of serinyl-tRNA^{Ser}, catalyzed by serinyl-tRNA^{Ser} synthetase. This modified aminoacyl-tRNA recognizes the codon, UGA. (a) Draw a possible structure for the selenocysteine residue. You may represent the tRNA as in text Figure 23.9. (b) What is unusual about the codon recognized by the aminoacyl-tRNA?

24

Recombinant DNA Technology

I. KEY TERMS

With the help of this study guide, your textbook, and class notes, you should be able to define and explain the significance of the following terms:

recombinant DNA technology

vector

genetic transformation

marker gene

selection

insertional inactivation

DNA library

cDNA

probe

expression vector

restriction fragment length polymorphism

polymerase chain reaction

recombinant DNA molecule

cloning

transfection

sticky end

screen

shuttle vector

cosmid

probing

chromosome walk

transgenic

II. EXERCISES

A. True–False

_____ **1.** Plasmids are circular, single-stranded, supercoiled molecules of DNA.

_____ **2.** Recombinant DNA molecules are inventions of man.

_____ **3.** A vector becomes a recombinant DNA molecule once insertion of a desired DNA fragment is accomplished.

_____ **4.** The normal flow of information goes from DNA to mRNA to protein, but it is possible to make DNA from the information provided by a specific mRNA molecule.

_____ **5.** Some vectors can replicate in either prokaryotic or eukaryotic cells.

_____ 6. If a DNA vector and the DNA fragment to be inserted have complementary sticky ends, a stable recombinant DNA molecule can be created by simply annealing the two types of DNA.

_____ 7. Some cells take up recombinant DNA molecules directly by a process called conjugation.

_____ 8. Insertion of a DNA fragment into the amp^R gene of the pBR322 plasmid would lead to transformants that are resistant to both ampicillin and tetracycline.

_____ 9. In the preparation of phage particles containing recombinant DNA molecules, the λ phage DNA is cleaved with a restriction endonuclease to yield λ arms and stuffer fragment(s). The stuffer fragment(s) is(are) purified so it(they) can be combined with genomic DNA to create the recombinant molecule.

_____ 10. The poly A tails of mature eukaryotic mRNA molecules are important to the formation of cDNA molecules using reverse transcriptase.

B. Short Answers

1. The direct uptake of foreign DNA by a host cell is called _____.

2. A collection of bacterial colonies, each containing a different duplex DNA segment inserted into a plasmid is called a/an _____.

3. _____ involves introduction of a normal gene into mutant cells of an individual in an effort to correct a genetic disease.

4. A _____ is a molecule that specifically recognizes a desired DNA segment in a large library.

5. Plasmids that have been modified to contain eukaryotic promoters, ribosome-binding sites, and transcription terminators are called _____, since they allow the expression of eukaryotic genes of recombinant DNA molecules in prokaryotic cells.

6. An animal or plant that has been engineered to contain stably-integrated foreign DNA is said to be _____.

7. Variations that occur in the pattern of restriction fragments, produced by the same restriction endonuclease, from the DNA of different individuals of the same species are known as _____.

8. Amplification of small amounts of DNA taken from a source such as a hair follicle is accomplished by a technique called the _____.

9. _____ is the technique by which a gene of a vector is rendered incapable of expression of the gene product due to the insertion of a foreign DNA fragment into the gene.

10. The introduction of recombinant DNA into a host cell by engineered phage particles occurs by a process called _____.

C. Problems

1. Name four different types of vectors that have used in recombinant DNA technology.

2. Without using enzymes, how could a cDNA-mRNA hybrid be easily treated in the laboratory to yield the cDNA as a single strand?

3. In what circumstance(s) is λ phage preferable to plasmids as a vector for genomic DNA?

4. (a) Why must λ phage DNA be engineered to be a suitable vector for recombinant DNA expression? (b) What is done to the phage DNA prior to using it to make recombinant DNA molecules?

5. In the production of cDNA using reverse transcriptase, why is DNA polymerase I used in the synthesis of duplex DNA rather than DNA polymerase III?

6. In what way are cDNA libraries preferable to genomic libraries?

7. Suppose a researcher was interested in determining whether or not particular amino acid residues of an enzyme were involved in the catalytic activity characteristic of the enzyme. What genetic engineering approach is now used in such studies?

8. The polymerase chain reaction (PCR) is used to amplify small amounts of DNA. Duplex DNA is heated to separate its strands for synthesis, since single DNA strands are required to serve as templates. (a) Since elevated temperatures denature not only DNA but also enzymes and other proteins, how is synthesis of the new strands of DNA accomplished? (b) Cite three or four circumstances in which the use of PCR allows investigators to obtain knowledge that would not be possible otherwise.

9. Problem 7. at the end of text Chapter 24 refers to mechanical shearing of DNA, which creates fragments with blunt ends. Usually, researchers want DNA fragments with sticky ends for construction of recombinant DNA molecules. Phage T4 DNA ligase is an enzyme that will catalyze the covalent joining of *any* two DNA molecules with blunt ends. How might one use this information to modify a DNA fragment that would contain a particular restriction site that would produce sticky ends when cleaved by the specific restriction endonuclease?

10. As previously mentioned, plasmid vectors have been engineered to allow expression of eukaryotic genes in prokaryotic cells. A hybrid promoter, p_{tac}, has been made that contains part of the *lac* promoter and part of the *trp* promoter. This hybrid promoter must be activated by isopropyl-β-thiogalactoside (IPTG) before it will drive expression of inserted foreign genes. If production of a eukaryotic protein is the desired end, why engineer a plasmid with a promoter that must be activated?

D. Additional Problems

1. Itakura and coworkers tried to produce the hormone somatostatin by recombinant DNA techniques. Initial attempts using the transformed bacterial cell were unsuccessful. The investigators then attached the somatostatin gene to the bacterial *lac* gene with a bridge between the two that consisted of the base sequence ATG in the template strand. Expression was successful. Why was the bridge sequence used? (Note that the somatostatin gene does not contain an ATG sequence.)

2. Radioactive probes, such as a DNA segment complementary to a single strand region of interest, are used to locate particular genes. Such complementary DNA (cDNA) sequences can be synthesized if the corresponding mRNA sequence is known. The problem is complicated, however, if only the sequence of the protein for which the mRNA is the blueprint is known. Explain.

3. With reference to the previous problem, how many different DNA probes would be necessary to find the DNA template strand that corresponds to the mRNA segment that gives rise to the peptide segment Trp-Phe-Lys-Glu-Met?

4. Write sequences of the DNA probes from the previous problem.

5. It is estimated that a 1-kb fragment of human DNA represents about 0.000031% of the human genome. What, then, is the size of the human genome?

6. How many clones must be screened, with a probability of 0.99 that a 10-kb fragment was located, from a genome of 144,000 kb?

Solutions

CHAPTER 2:

A. True–False

1. _False._ The highly electronegative oxygen pulls the shared electrons closer to the oxygen than the hydrogen, giving rise to the polar bond. The electronegativity of hydrogen is considerably less than that of oxygen.

2. _True._ Because the H–O–H bond angle is 104.5°, the two polar O–H bonds result in a center of negative charge for the molecule that does not coincide with the center of positive charge, as indicated in text Figure 2.2.

3. _True._ Potential hydrogen bonds can involve the two hydrogen atoms of H_2O and the two non-bonding electron pairs on the oxygen atom. This possibility is illustrated for ice in text Figure 2.5. Some sources indicate, however, that an average of 3.4 hydrogen bonds form in liquid water.

4. _True._ Electrolytes (like NaCl) are polar compounds that dissociate into their anions and cations, each surrounded by water molecules which form a solvation sphere. Nonelectrolytes like glucose and ethanol also become hydrated by water molecules due to hydrogen bond formation between water and the polar —OH groups of the solvated molecules.

5. _False._ The osmotic pressure depends on the total molar concentration of solute, not on the chemical nature of the solute(s).

6. _True._ The dipoles thus produced are short lived and the strengths of the induced interactions are low.

7. _True._ At pH 3.0, however, the $[H^\oplus] > [OH^\ominus]$.

8. _False._ An acid is weak if it dissociates less than 100% in water.

9. _True._ Use the pK_a for lactic acid from text Table 2.4 and employ the Henderson–Hasselbalch equation. When $[A^\ominus]/[HA] = 1$, the $\log [A^\ominus]/[HA] = 0$, and $pH = pK_a$ of the weak acid.

10. _True._

11. _False._ It is called a solvation sphere.

B. Short Answers

1. Electronegativity

2. Chaotropic

3. Amphipathic

4. micelles

5. solvation sphere

6. van der Waals forces

7. M^2 or (moles/liter)^2

8. weak

9. M or moles/liter

10. buffer

11. acidosis

12. electrolyte

C. Problems

1. HCl, a strong acid, ionizes 100%.

$$HCl \rightarrow H^\oplus + Cl^\ominus$$

Therefore, $[H^\oplus] = [Cl^\ominus] = 1.00 \times 10^{-2}$.

$$pH = -\log (1.00 \times 10^{-2}) = 2.00.$$

Acetic acid is weak. It does not ionize 100%.

$$CH_3COOH \rightleftharpoons H^{\oplus} + CH_3COO^{\ominus}$$

Initial:	0.0100 M	0	0
At Equilibrium:	0.0100 – x	x	x

where x represents the moles/liter of acetic acid which ionize and the moles/liter of $H^{\oplus}$ and of $CH_3COO^{\ominus}$ present at equilibrium due to that ionization. You can neglect x in the term $0.0100 - x$ when the acid K_a < about 10^{-4} and when the original concentration of the weak acid is > about 10^{-2} M. Making this assumption here, the equilibrium concentration of acetic acid is:

$$K_a = \frac{[H^{\oplus}][CH_3COO^{\ominus}]}{[CH_3COOH]}$$

From text Table 2.4, the K_a for acetic acid is = 1.76×10^{-5}.
Therefore

$$1.76 \times 10^{-5} = \frac{(x)(x)}{0.0100}$$

$$x^2 = (0.0100)(1.76 \times 10^{-5}) = 1.76 \times 10^{-7}$$

$$x = [H^{\oplus}] = 4.20 \times 10^{-4}$$

$$pH = -\log(4.20 \times 10^{-4}) = 3.38.$$

The HCl puts more $H^{\oplus}$ ions in solution since it is 100% ionized. Acetic acid is less than 100% ionized and releases fewer $H^{\oplus}$ ions, which results in a higher pH.

2. HNO_3 is a strong acid, so $[H^{\oplus}] = 2.50 \times 10^{-3}$.

$$pH = -\log 2.50 \times 10^{-3} = 2.60.$$

$$K_w = [H^{\oplus}][OH^{\ominus}] = 1.00 \times 10^{-14}; \text{ therefore,}$$

$$[OH^{\ominus}] = (1.00 \times 10^{-14})/(2.50 \times 10^{-3}) = 4.00 \times 10^{-12}.$$

3. The K_a for acetic acid is 1.76×10^{-5}. The pK_a is 4.75. Note that, since the pH is lower than the pK_a, one should expect a greater acid than conjugate base concentration (i.e., $[CH_3COO^{\ominus}]$ < $[CH_3COOH]$).

Use the Henderson–Hasselbalch equation:

$\log [CH_3COO^{\ominus}]/[CH_3COOH] = pH - pK_a$

$[CH_3COO^{\ominus}]/[CH_3COOH] = \text{antilog} [pH - pK_a]$

$[CH_3COO^{\ominus}]/[CH_3COOH] = \text{antilog} [4.00 - 4.75]$

$[CH3COO^{\ominus}]/[CH_3COOH] = 0.178$. Thus, the ratio of acetate to acetic acid is 0.178 to 1 or 1 to 5.62.

4. If 23 kJ represents the strength of each of the four hydrogen bonds present in ice, 78.2 kJ would indicate that fewer than four hydrogen bonds are present in liquid water. The average number of hydrogen bonds is:

$$\frac{78.2 \text{ kJ}}{\text{molecule}} \times \frac{1 \text{ hydrogen bond}}{23 \text{ kJ}} = \frac{3.4 \text{ hydrogen bonds}}{\text{molecule}}$$

5. $H_3O^{\oplus}$ and H_2O are a conjugate acid–base pair, but so are H_2O and $OH^{\ominus}$. Water is amphoteric, which means it can act both as an acid and as a base.

6. If you jump to your calculator immediately without thinking about this problem, you will probably answer incorrectly. The pH is not 8. That is a basic pH. In this dilute solution, the $H^{\oplus}$ from the ionization of water is the major contributor to the pH. $[H^{\oplus}]$ from water is 1.0×10^{-7} and that from the HCl is 10^{-8} or 0.1 x 10^{-7}. Use the sum of the two concentrations, 1.1 x 10^{-7}, to determine the pH.

$$pH = -\log 1.1 \times 10^{-7} = 6.958 = 7.0, \text{ to two significant figures.}$$

Not very acidic, is it?

7. For weak acids at low concentrations,

$$[H^\oplus] = (K_a\,[HA])^{1/2}$$
$$[H^\oplus] = [(1.38 \times 10^{-4})\,(0.0112)]^{1/2} = 1.24 \times 10^{-3}.$$
$$pH = -\log(1.24 \times 10^{-3}) = 2.91.$$

8. At this point in the titration, half of the lactic acid molecules have had their protons removed (titrated) by $OH^\ominus$ ions of the added base, forming water and an equal amount of lactate ion. With one–half the lactic acid still present in protonated form, there are equal amounts of a weak acid (lactic) and its conjugate base (lactate ion). This is a buffer system. The pH is the same as the pK_a of the acid, since the Henderson-Hasselbalch equation term $\log[A^\ominus]/[HA] = 0$. Therefore,

$$pH = pK_a + 0 = 3.86.$$

9.

$$\begin{aligned} pH &= pK_a + \log([A^\ominus]/[HA]) \\ &= 3.19 + \log[(0.0400)/(0.0450)] \\ &= 3.19 - 0.05 = 3.14. \end{aligned}$$

10. H_2CO_3 forms rapidly by reaction of dissolved carbon dioxide with water.

$$H_2O + CO_2 \rightleftharpoons H_2CO_3$$

As H_2CO_3 is used, more is formed from the carbon dioxide carried in the blood, thus assuring an adequate supply.

D. Additional Problems

1. If acetic were a strong acid, the $[H^\oplus]$ would be 1.00×10^{-2}. However, $[H^\oplus]$ was found to be 4.20×10^{-4}. Thus, the % ionization $= (4.20 \times 10^{-4}/1.00 \times 10^{-2}) \times 100 = 4.20\%$.

2.

$$\begin{aligned} pH &= pK_a + \log([A^\ominus]/[HA]); \text{ and,} \\ pK_a &= pH - \log([A^\ominus/[HA]). \\ &\text{If blood pH} = 7.4, \\ pK_a &= 7.4 - \log(20/1) \\ &= 7.40 - 1.30 \\ &= 6.10. \end{aligned}$$

Table 2.4 of your text gives a value of 6.4. Note: be aware that different texts use different values for the pK_a of carbonic acid. Ask your instructor which one he/she thinks is most appropriate.

3. If $K_w = 1.4 \times 10^{-14}$ M^2, $[H^\oplus] = (1.4 \times 10^{-14})^{1/2} = 1.18 \times 10^{-7}$ and pH $= -\log(1.18 \times 10^{-7}) = 6.93$. The solution is neutral because it contains equal quantities of $H^\oplus$ and $OH^\ominus$, since one of each is formed when an H_2O molecule ionizes.

$$H_2O \rightleftharpoons H^\oplus + OH^\ominus$$

This illustrates that the familiar pH scale is based on a specific temperature (25°C).

4. In the Henderson–Hasselbalch equation, the quantities of $A^\ominus$ and HA are expressed as moles/liter. Since both species are in the same solution and thus have the same volume, the volumes in the $[A^\ominus]/[HA]$ ratio cancel. A ratio of the moles of $A^\ominus$/HA remains.

$$\text{Moles } A^\ominus\,(HPO_4^{2\ominus}) = 0.4000\text{L} \times 0.12\text{ mol/L} = 0.048.$$
$$\text{Moles HA }(H_2PO_4^\ominus) = 0.6000\text{L} \times 0.075\text{ mol/L} = 0.045.$$
$$\begin{aligned} pH &= 7.21 + \log[(0.048)/(0.045)] \\ &= 7.21 + 0.028 = 7.238 = 7.24. \end{aligned}$$

5. Assume that all dissolved CO_2 forms H_2CO_3: $H_2O + CO_2 \rightarrow H_2CO_3$.

$$\text{If pH} = 5.68, [H^{\oplus}] = 2.09 \times 10^{-6}. \text{ Also,}$$
$$[H^{\oplus}] = ([HA]\, K_a)^{1/2} \text{ or } [H^{\oplus}]^2 = [HA]\, K_a. \text{ Then,}$$
$$[HA] = [H^{\oplus}]^2/K_a = (2.09 \times 10^{-6})^2/(4.3 \times 10^{-7})$$
$$= 1.02 \times 10^{-5} \text{ M.}$$

CHAPTER 3:

A. True–False

1. <u>True.</u>   The α-amino group ($pK_a = 9.9$) of all aspartic acid molecules will be protonated, $-NH_3^{\oplus}$, as will half the α-carboxyl groups, -COOH. The other half will be ionized ($-COO^{\ominus}$), since the pK_a for that group is 2.0. Since the pK_a for the side chain carboxyl group is 3.9, few of those groups will be ionized and the majority of the molecules will then carry a net positive charge.

2. <u>True.</u>   Use the Henderson-Hasselbalch equation to determine the amount of conjugate base to acid form for each of the three ionizable groups at pH 10. On the basis of 100 lysine molecules:

$$\alpha\text{-amine group:} \quad 11\ NH_3^{\oplus}, 89\ NH_2$$
$$\alpha\text{-carboxyl group:} \quad 100\ COO^{\ominus}, 0\ COOH$$
$$\text{amine of R-group:} \quad 76\ NH_3^{\oplus}, 24\ NH_2$$

The probability of a form with three separate charges is:

$$p = 11/100 \times 100/100 \times 76/100 = 0.0836 = 0.084$$

So about 8.4% of the molecules have three separate charges, not the predominant form.

3. <u>False.</u>   Only seven of the 20 amino acids are classified as highly hydrophilic.

4. <u>True.</u>

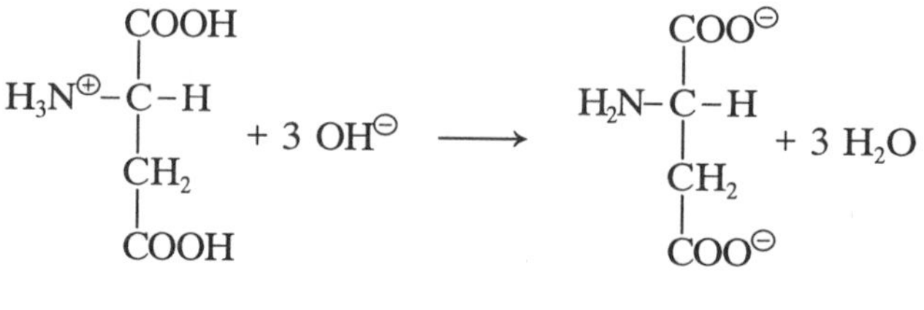

5. <u>False.</u>   Glycine is not. It has no chiral center since its R-group is an H atom.

6. <u>True.</u> $pI = (pK_{a1} + pK_{a2})/2$.

7. <u>False.</u>   Direction of rotation of polarized light must be determined empirically. L- refers to the structural similarity to L-glyceraldehyde.

8. <u>False.</u>   The solution would show no optical activity because the equal and opposite optical activities of the D- and L-valine components would cancel. Equal amounts of two enantiomers constitutes a racemic mixture.

9. <u>False.</u>   It is a pentapeptide. It is the number of amino acid residues that dictates the prefix used. Five residues would involve four peptide bonds, as follows: A-B-C-D-E (the dashes (-) represent peptide bonds).

10. <u>False.</u>   Cysteine is the only exception. L-cysteine is R-cysteine because the side chain, $-CH_2SH$, has a higher priority than the carboxyl group, -COOH. See Figure 3.1.

11. <u>False.</u>   A pH of 5.7 is the isoelectric point for methionine, so methionine will not migrate. Aspartic acid will have a net negative charge at pH 5.7 and will migrate toward the anode.

R—designation

Figure 3.1 The *RS* Designation of L-cysteine

12. <u>True.</u> SDS denatures proteins and associates with protein molecules giving all molecules a negative charge proportional to their length. Thus the charge/mass ratios are the same for all SDS-proteins. Migration rates depend on size and charge/mass ratios. Since these proteins have the same charge/mass ratio, they will migrate according to size only.

B. Short Answers

1. eluted

2. zwitterion

3. isoelectric point

4. disulfide bridge

5. hydropathy

6. microenvironment

7. enantiomers

8. primary structure

9. Edman degradation procedure; sequenator

10. N-terminus; C-terminus

C. Problems

1. The pK_a for the α-carboxyl group of alanine is 2.4. Use the Henderson-Hasselbalch equation to determine the ratio of conjugate base groups (-COO$^\ominus$) to acid groups (-COOH).

$$pH = pK_a + \log [A^\ominus]/[HA]; \text{ let } x = [A^\ominus]/[HA].$$
$$\log x = pH - pK_a = 3.0 - 2.4 = 0.6$$
$$x = \text{antilog } 0.6 = 3.98.$$

So, there are 4 -COO$^\ominus$ groups for every 1 -COOH group (5 total groups).

Per 100 alanine molecules, there are $4 \times 20 = 80$ ionized carboxyl groups and $1 \times 20 = 20$ unionized groups.

2.

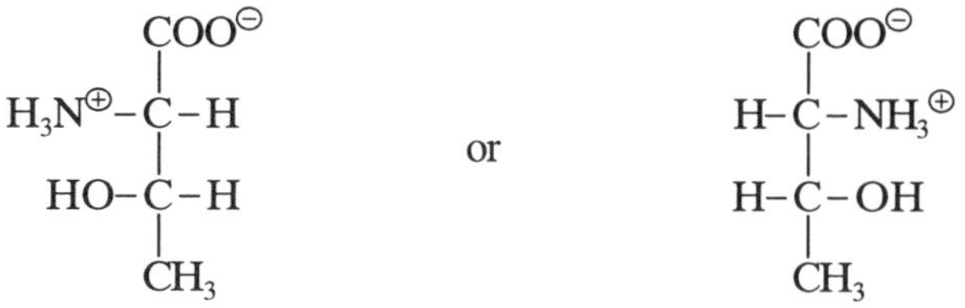

Note that, although these forms are enantiomers of *each other*, each is a diastereomer of the form given in the question.

3. Although glycine has a relatively low molecular weight (75 g/mol), it exists in a salt form (zwitterionic form) and has properties similar to other salts, namely stability of crystals to relatively high temperature. Amino acids typically decompose over a relatively high temperature range.

4.

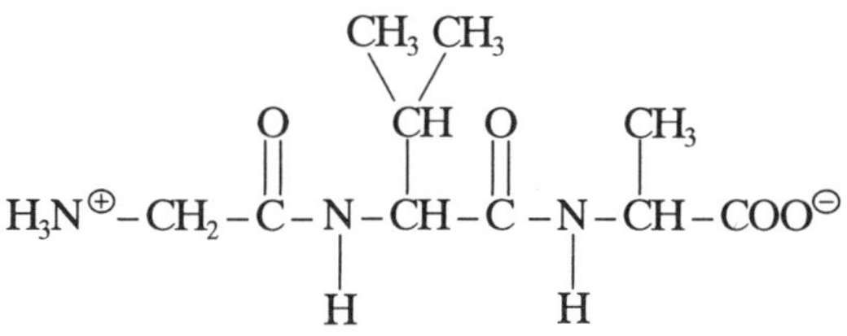

5. See text Table 3.2 for pK_a values of serine.

a) $H_3N^{\oplus}-CH-COOH$
    $\quad\quad\quad\quad |$
    $\quad\quad\quad CH_2$
    $\quad\quad\quad\quad |$
    $\quad\quad\quad OH$

b) $H_3N^{\oplus}-CH-COO^{\ominus}$
    $\quad\quad\quad\quad |$
    $\quad\quad\quad CH_2$
    $\quad\quad\quad\quad |$
    $\quad\quad\quad OH$

c) $H_2N-CH-COO^{\ominus}$
    $\quad\quad\quad |$
    $\quad\quad CH_2$
    $\quad\quad\quad |$
    $\quad\quad OH$

Note: These are the major forms in each population.

6. (a) At pH 3.0, virtually all of the N-terminal amino groups will be protonated ($-NH_3^{\oplus}$). By use of the Henderson-Hasselbalch equation, you can determine that only 11 of every 100 molecules of aspartame will have the carboxyl group of the aspartate side chain ionized. No other groups are capable of ionizing since the C-terminus exists as a methyl ester. Therefore, the predominate ionic form of aspartame at pH 3.0 is positively charged and will migrate toward the negative electrode, the cathode.

(b) Use the pK_a values in text Table 3.2 for the α-amino group and the side chain carboxyl group of aspartate to calculate the isoelectric point of aspartame. It is 6.9, so the aspartame sample should not migrate in an electric field at pH 6.9. Note again that no other groups can contribute to the ionic form of the molecules at this pH.

7.

CHO
$\quad |$
HO$-$C$-$H
$\quad |$
H$-$C$-$NH$_2$
$\quad |$
CH$_2$OH

enantiomer

CHO
$\quad |$
H$-$C$-$OH
$\quad |$
H$-$C$-$NH$_2$
$\quad |$
CH$_2$OH

or

CHO
$\quad |$
HO$-$C$-$H
$\quad |$
H$_2$N$-$C$-$H
$\quad |$
CH$_2$OH

diastereomers

8. You will find pK_a values for these and the other amino acids in Table 3.2 of your text.

Glycine: pI = (2.4 + 9.8)/2 = 6.1

Lysine: pI = (9.1 + 10.5)/2 = 9.8

Aspartic Acid: pI = (2.0 + 3.9)/2 = 3.0

Note that, for lysine, the two like (basic) groups only were averaged because the pK_a values of these groups lie equidistant on either side of the pH at which there exists a species with a net zero charge. The two like groups of aspartate are the two acidic groups.

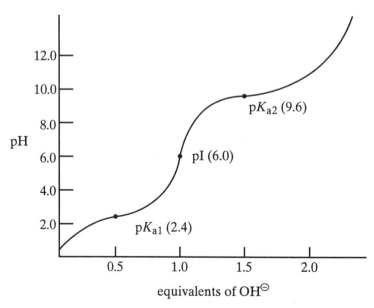

Figure 3.2 Titration Curve for an Amino Acid.

9. Figure 3.2 shows the titration curve of an amino acid with a pK_{a1} of 2.4 and a pK_{a2} of 9.6. Note that one equivalent of $OH^{\ominus}$ titrates one equivalent of -COOH, converting it all to the $-COO^{\ominus}$ form. All the amino groups are still protonated ($-NH_3^{\oplus}$). This corresponds to the isoelectric point. The addition of 0.5 equivalent of $OH^{\ominus}$ titrates one-half of the total -COOH groups resulting in equal numbers of -COOH and $-COO^{\ominus}$ groups. From the Henderson-Hasselbalch equation, the pH at this point is equal to the pK_a of the -COOH group. Similarly, when 1.5 equivalents of $OH^{\ominus}$ have been added, half of the protonated amino groups have been titrated and there are thus equal amounts of $-NH_2$ and $-NH_3^{\oplus}$ forms. The pH at that point is the pK_a of the amino group.

10.

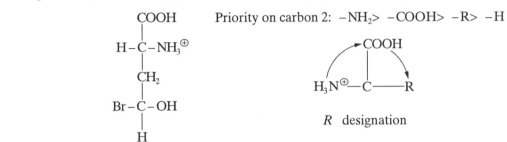

Priority on carbon 2: $-NH_2> -COOH> -R> -H$

R designation

Priority on carbon 4: $-Br> -OH> -R> -H$

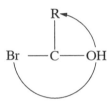

S designation

(2*R*, 4*S*)-2-amino-3-bromo-3-hydroxybutanoic acid

D. Additional Problems

1. For the α-amino group, $pK_a = 9.5$. Use the Henderson-Hasselbalch equation:

$pH = pK_a + \log [-NH_2]/[-NH_3^{\oplus}]$; let x = $[-NH_2]/[NH_3^{\oplus}]$.

$\log x = pH - pK_a = 4.00 - 9.5 = -5.5$

$[NH_2]/[NH_3^{\oplus}]$ = antilog $-5.5 = 3.2 \times 10^{-6}$ to 1.

The α-carboxyl group $pK_a = 2.1$.

Let x = $[-COO^{\ominus}]/[-COOH]$

$\log x = 4.00 - 2.1 = 1.9$.

$[COO^{\ominus}]/[COOH]$ = antilog $1.9 = 79$ to 1.

The side chain carboxyl group $pK_a = 4.1$.

Let x = $[-COO^{\ominus}]/[-COOH]$

$\log x = 4.00 - 4.1 = -0.1$.

$[COO^{\ominus}]/[COOH]$ = antilog $-0.1 = 0.79$ to 1.

2. There are 20 possibilities for each of 15 residue sites, or 20^{15}.

$\log$ answer = $\log 20^{15} = 15(\log 20) = 19.52$;

answer = antilog $19.52 = 3.28 \times 10^{19}$.

3. Since the isoelectric pH of glycine is 6.1, it has very little tendency to move in the electric field. Having a <u>slight</u> excess of *positive* charge (since the pH is not quite at the isoelectric pH), it may travel slightly toward the negative electrode. The total time the electric field is applied is also a factor in extent of travel. Lysine has a net *positive* charge since its $-NH_3^{\oplus}$ groups tend not to release protons appreciably before pH 7-8. Lysine will migrate appreciably toward the negative electrode. Aspartate has a net *negative* charge since both -COOH groups are fully ionized at pH 6.0. It will migrate toward the positive electrode.

4. The dipeptide composed of alanine and aspartic acid could have either of these primary structures:

$$H_3N^{\oplus}—Ala—Asp—COOH \quad or \quad H_3N^{\oplus}—Asp—Ala—COOH$$
$$\underset{\displaystyle COOH}{|} \qquad\qquad\qquad \underset{\displaystyle COOH}{|}$$

Each dipeptide has three groups that can donate protons. Thus, one mmole of dipeptide requires three mmole of base.

$$(30. \text{ ml of } 0.1 \text{ M NaOH} = 3 \text{ mmole NaOH}).$$

The dipeptide composed of alanine and glycine could have either of these primary structures:

$$H_3N^{\oplus}—Ala—Gly—COOH \quad or \quad H_3N^{\oplus}—Gly—Ala—COOH$$

Each dipeptide has only two groups to react with base and would thus consume only 2 mmole of NaOH. The dipeptide which was titrated must then be the one composed of alanine and aspartic acid (the order is unknown).

5. Note that there might be faster and less tedious ways of sequencing such a small peptide (perhaps with a sequenator). This problem, however, is designed to help you understand how larger polypeptides might have to be segmented by various cleaving agents in order to deduce their primary structures. To solve this problem, it may be helpful to make blanks for each of the amino acid residues from the N-terminus to the C-terminus and fill in the identities of each residue as deduced from the data.

Equimolar amounts of Arg, Glu, Gly, Lys, Met, and Phe indicate that this is a hexapeptide.

$$\text{N-term } [(\ )\text{-}(\ )\text{-}(\ )\text{-}(\ )\text{-}(\ )\text{-}(\ )] \text{ C-term}$$

(a) Since trypsin cleaves on the carbonyl side of lysine and arginine, the fact that arginine is found alone and the rest of the peptide left intact suggests that arginine is N-terminus and lysine is C-terminus. If lysine were in the interior of the peptide, another cleavage would have occurred. So,

$$\text{N-term } [Arg\text{-}(\ )\text{-}(\ )\text{-}(\ )\text{-}Lys] \text{ C-term}$$

(b) Cyanogen bromide cleaves on the carbonyl side of methionine residues. Since this resulted in two tripeptides, the methionine must be the third residue from the N-terminus. You should note that the methionine residue would have been converted to homoserine lactone as indicated in your text. Thus,

$$\text{N-term } [Arg\text{-}(\ )\text{-}Met\text{-}(\ )\text{-}(\ )\text{-}Lys] \text{ C-term}$$

(c) The *S. aureus* V8 protease cleaves on the carbonyl side of aspartate and glutamate residues. This action left free lysine and a pentapeptide. The lysine had to be on the carbonyl side of glutamate for this result to have occurred. This confirms that lysine is the C-terminal residue and that glutamic acid is the second residue from the C-terminus. Therefore,

$$\text{N-term } [Arg\text{-}(\ )\text{-}Met\text{-}(\ )\text{-}Glu\text{-}Lys] \text{ C-term}$$

(d) Chymotrypsin cleaves on the carbonyl side of aromatic residues like phenylalanine. Since there are only two positions left available (2 and 4) in the hexapeptide, consider the possible results if phenylalanine were in each position. In position 2, cleavage would still give a dipeptide and a tetrapeptide, but the UV absorbance (260 nm) would be observed in the dipeptide and not the tetrapeptide. Therefore, phenylalanine must be in position 4. The only residue left is glycine and the only position left is position 2. The total primary structure is:

$$Arg\text{-}Gly\text{-}Met\text{-}Phe\text{-}Glu\text{-}Lys$$

CHAPTER 4

A. True–False

1. <u>False.</u> Hydrogen bonds stabilize areas of secondary structure.

2. <u>True.</u> The *trans* conformation is spatially more favorable than the *cis* conformation. Rare exceptions occur, usually at peptide bonds involving the amide nitrogen of proline.

3. <u>True.</u> It is the native conformation that the protein normally has in its biological environment.

4. <u>False.</u> The capacities to bind oxygen are significantly different (due primarily to the lack of quaternary structure in myoglobin), but the structures of the protein chains are very similar.

5. <u>False.</u> The pitch is the distance along the α-helix that constitutes one turn of the helix and involves about 3.6 amino acid residues.

6. <u>False.</u>   An isolated heme in solution does not bind oxygen. Instead, the Fe^{2+} ion of heme is oxidized to Fe^{3+}. This does not occur in the hemoglobin or myoglobin molecules due to the microenvironment of the heme group in these proteins.

7. <u>False.</u>   They are called motifs. Domains are independently-folded regions and may consist of combinations of motifs.

8. <u>True.</u>   This allows interaction of these water-loving groups with the water molecules that surround the protein.

9. <u>False.</u>   Normally, no covalent bonds are broken. Activity is lost, however, due to loss of the native tertiary structure, which is maintained by non-covalent interactions.

10. <u>True.</u>   Salt bridges (positive ion attracted to negative ion), hydrogen bonding, hydrophobic forces, and other interactions help maintain the association of the monomers of an oligomeric protein.

11. <u>False.</u>   Low pH (high $[H^+]$) causes a lowered oxygen affinity.

B. Short Answers

1. the Bohr effect

2. the α-helix

3. extremes of pH; the use of 8 M urea

4. a random coil

5. oligomeric proteins

6. the Fe^{2+} ion of the heme group

7. multienzyme complex

8. renaturation

9. positive cooperativity of binding

10. 2,3-*bis*phosphoglycerate (BPG)

11. Molecular chaperones

12. right-handed helix (α-helix or 3_{10} helix)

13. antigens; antibodies

C. Problems

1. The advance along the axis in an α-helix is about 0.15 nm per residue, so the segment is 0.15 nm/residue × 20 residues = 3.0 nm in length.

2. Pauling showed that there are about 3.6 residues per turn, so 20 residues × 1 turn/3.6 residues = 5.6 turns.

3. Carbonyl oxygens of peptide bonds are hydrogen bonded to the α-amino nitrogen of the fourth residue "ahead" (toward the C-terminus of the chain). Exceptions are that the first four α-amino nitrogens and the last four carbonyl oxygens of the α-helical segment are not involved in hydrogen bonding in the α-helix. The hydrogen bonds are roughly parallel to the helix axis.

4. Draw a linear protein segment of 10 residues. Label the R groups R_n, R_{n+1}, and so on, through R_{n+9}. Draw hydrogen bonds between the carbonyl oxygen of residue n to the amide hydrogen of residue $n+4$ as indicated on page 88 of your text. This procedure will result in a variance with what is stated in parentheses in your text (page 88, third line) because we have assumed that all atoms of the 10 residues are "in" the helix. With 10 residues, there are 10 carbonyl oxygens and 10 α-amino nitrogens. It takes one of each to make a hydrogen bond, so there is the potential for 10 hydrogen bonds. However, four α-amino nitrogens at the N-terminal end of the helix and four carbonyl oxygens at the C-terminal end of the helix are not involved in hydrogen bonding. Thus, there are 10 – 4 = 6 hydrogen bonds. That represents 60% of the possible hydrogen bonds. Note: This percentage will change with the length of the chain. For a helical segment of 20 residues, there are 16 hydrogen bonds, representing 80% of possible hydrogen bonds. Does it appear reasonable that the number of hydrogen bonds in any α-helix will be n – 4, where n is the number of residues in the helix?

5. The first three residues, if able to participate at all, will contribute less than one turn (3 residues × 1 turn/3.6 residues = 0.8 turns or 3 residues × 0.15 nm/residue = 0.45 nm). Proline is generally a helix breaker and will usually form a bend. The next six residues could form 6 residues × 1 turn/3.6 residues = 1.7 turns or 6 residues × 0.15 nm/residue = 0.90 nm of helix before the next proline bends the chain.

6. Normal collagen contains hydroxyproline and hydroxylysine residues that are formed by hydroxylation of proline and lysine, respectively. Proline is hydroxylated by the action of prolyl hydroxylase, which requires vitamin C for its

activity. Without vitamin C, proper amounts of hydroxyproline and hydroxylysine are apparently not produced and collagen cannot form fibers properly. This must result in abnormal collagen that is unable to perform the normal structural functions.

7. RNase A contains four disulfide bonds in its native, biologically-active form that are broken by the action of the 2-mercaptoethanol. For the denatured, inactive protein to regain biological activity, it must regain its native conformation with the four disulfide bonds formed between the original (native) cysteine partners. In the presence of urea, the protein is prevented from properly refolding and reoxidation produces random disulfide bond formation for about 99% of the molecules. One may conclude that the protein chain must be able to fold into its native conformation, which can occur with both urea and 2-mercaptoethanol removed, so that the native disulfide bonds can reform and biological activity be regained.

8. Figure 4.49 in your text shows that hemoglobin is roughly 45% saturated ($Y = .45$) with oxygen at pH 7.2 when pO_2 = 30 torr. At pH 7.6, the degree of oxygen saturation is increased to about 73%. The ratio of oxygen saturation, then, is 0.73/0.45 = 1.6. Therefore, at pO_2 of 30 torr, a change of pH from 7.2 to 7.6 allows hemoglobin to hold 1.6 times as much oxygen.

9. (a) $R-NH_2 + CO_2 \rightleftharpoons R-NH-\overset{\displaystyle |}{\underset{\displaystyle O^{\ominus}}{C}}=O + H^{\oplus}$

(b) $HCO_3^{\ominus} + H^{\oplus} \rightleftharpoons H_2CO_3 \rightleftharpoons H_2O + CO_2$

Note that although these reactions are reversible, product (CO_2) formation is favored in the lungs where CO_2 concentration is low.

(c) Since only four N-terminal amino groups per hemoglobin molecule are potentially available for carbamate adduct formation, it would seem that this might not be an important means of CO_2 transportation. Thomas Devlin, in *The Textbook of Biochemistry with Clinical Correlations*, 3rd Edition, 1992, page 1036, estimates the major forms of carbon dioxide transport as:

$HCO_3^{\ominus}$	78%
dissolved CO_2	9%
carbamino hemoglobin	13%

10. An immunoassay employs antigen-antibody reactions for the determination of chemical substances. The specificity of the antibody for a particular antigen (such as a protein) makes the assay for that antigen possible. The enzyme-linked immunosorbant assay or ELISA is used to detect small amounts of specific proteins. For example, it is the basis for a pregnancy test in which the placental hormone chorionic gonadotropin is detected in the urine of the female within a few days after conception. (Voet & Voet, *Biochemistry*, 1990, pages 77, 78.)

D. Additional Problems

1. A carbon-carbon bond requires about 345 kJ/mole to break. (Holtzclaw, *General Chemistry*, 9th Edition, 1991, page 194.) A hydrogen bond requires about 2-20 kJ/mol, or an average of 9 kJ/mol for its disruption. (Figure 2.13 of your text.) Therefore, 345/9 = 38 hydrogen bonds to provide the energy equivalent of one carbon-carbon covalent bond.

2. (a) Six disulfide bridges involve 12 cysteine residues. Random reformation of the native disulfide bridges would be found by the probability calculation:

$$P = {}^1/_{11} \times {}^1/_9 \times {}^1/_7 \times {}^1/_5 \times {}^1/_3 \times {}^1/_1 = {}^1/_{10,400}$$

This probability is about 0.01%. The fact that there was 8% recovery suggests that something other than random chance played a role.

(b) Trypsin is synthesized in the body in the form of the inactive zymogen, trypsinogen. Active trypsin is formed by the removal of a segment of residues from the N-terminus of trypsinogen. Thus, the primary structure of the zymogen form, which determined the tertiary structure, is not the same as in that remaining in trypsin. Hence, the original tertiary structure of trypsin might not be expected to be regained.

3. Myoglobin contains 153 residues, 121 of which (Voet & Voet, page 219) are involved in the eight regions of α-helix, so that the potential for 121 hydrogen bonds exists. Since each α-helix will have four α-amino nitrogens (and four carbonyl oxygens) that do not form peptide bonds, there will be $4 \times 8 = 32$ fewer hydrogen bonds than theoretically possible. So, the number of hydrogen bonds would be $121 - 32 = 89$. At 9 kJ/mol, there would be 89 hydrogen bonds $\times$ (9 kJ/mol)/hydrogen bond= 801 kJ/mol stabilization.

4. The polypeptide chains in α-keratin are in the α-helix form. If you consider stretching out an α-helix until it is in the elongated β-sheet form, you can calculate the percent extension. The distance of one residue in an α-helix is 0.15 nm. In a pleated sheet, it is 0.32-0.34 nm. The percent of extension is then:

$$(0.33 - 0.15)\text{nm} \times 100/0.15 \text{ nm} = 120\%$$

5. For the dissociation:

$$MbO_2 \rightleftharpoons Mb + O_2$$

The dissociation constant, K_d, is:

$$K_d = \frac{[Mb]\, pO_2}{[MbO_2]}$$

At half-dissociation, $[Mb] = [MbO_2]$ and $pO_2 = P_{50}$. Placing these values in the dissociation constant expression gives: $Kd = P_{50}$. From this,

$$P_{50} = \frac{[Mb]\, pO_2}{[MbO_2]} \text{ and,}$$

$$\frac{[Mb]}{[MbO_2]} = \frac{P_{50}}{pO_2} = \frac{2.8 \text{ torr}}{6 \text{ torr}} = \frac{2.8}{6}$$

$$\%[Mb] = \frac{[Mb]}{[Mb]+[MbO_2]} \times 100 = \frac{2.8 \times 100}{2.8 + 6} = 32\%.$$

CHAPTER 5

A. True–False

1. <u>False.</u> Recent evidence indicates that some RNA molecules called ribozymes also possess catalytic activity.

2. <u>False.</u> Many enzymes require no cofactors at all.

3. <u>True.</u> Enzymes are usually rather specific in terms of the substrates they recognize. In many cases, the D-form of a sugar is recognized but not the L- form (its mirror image or enantiomer).

4. <u>False.</u> The substrate must fit specifically in the active site of the enzyme.

5. <u>False.</u> Complexes of substrate analogs and enzymes have been observed by x-ray crystallography.

6. <u>False.</u> This is a description of a ping-pong mechanism.

7. <u>False.</u> Neither can be precisely determined from such a plot. Use of a Lineweaver-Burk plot gives better approximations.

8. <u>False.</u> The lower the K_m, the greater the affinity of the enzyme for its substrate.

9. <u>True.</u> As a consequence, intracellular enzymatic rates are sensitive to small changes in substrate concentrations. See Text Figure 5.5.

10. <u>True.</u>

11. <u>True.</u> These enzymes have one or more modulator binding sites (perhaps for both positive and negative modulators) as well as the active site. Almost all regulatory enzymes discovered to date are oligomeric.

B. Short Answers

1. active site

2. enzyme-substrate complex

3. isomerases

4. catalytic constant, k_{cat} (or turnover number)

5. specificity constant; specificity of the enzyme for its substrate(s)

6. covalent modification

7. competitive

8. noncompetitive

9. uncompetitive

10. feedback inhibition

11. regulatory sites

C. Problems

1. Figure 5.1 in your text illustrates the solution to this problem. More product is formed per unit time in response to increasing amounts of enzyme. The system should be buffered and temperature kept constant at a pH and temperature that produces optimum reaction velocity.

2.

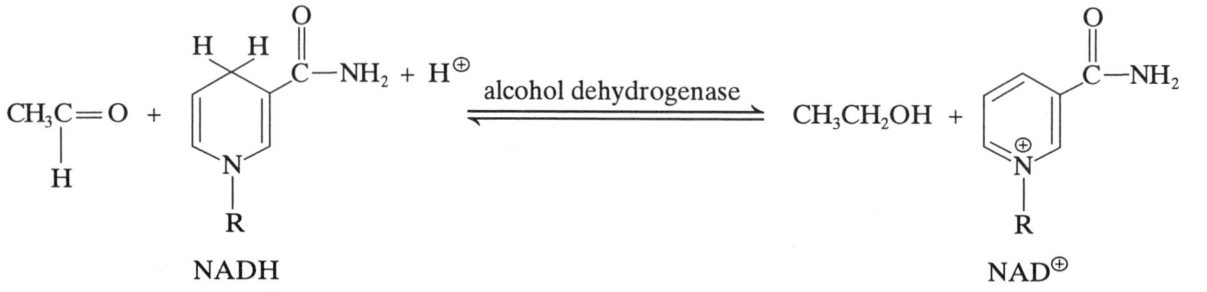

3. Since NADH absorbs at 340 nm, changes in the NADH concentration can be estimated at this wavelength. As a starting material, NADH would be present in large quantity as the reaction begins. It is more difficult to detect small changes in a large amount of a substance than it is to detect the appearance of a substance (NADH) when none was previously present, as would be the case for the reverse reaction.

4. Using the Michaelis-Menten equation:

$$v_0 = \frac{V_{max}[S]}{K_m + [S]};\qquad \text{let } [S] = 2K_m. \text{ Then}$$

$$v_0 = \frac{V_{max}(2K_m)}{K_m + 2K_m};\qquad v_0 = \frac{V_{max}(2K_m)}{3K_m};$$

$$v_0 = \tfrac{2}{3}V_{max}.$$

Similarly, for $[S] = 3K_m, v_0 = \tfrac{3}{4}V_{max}.$ And, for $[S] = 4K_m, v_0 = \tfrac{4}{5}V_{max}.$

5. Competitive inhibition is said to be overcome by an excess of substrate. Since the substrate and a competitive inhibitor compete with each other for the active site of a particular enzyme, a large excess of substrate insures that substrate molecules will bind to the active site nearly to the exclusion of competitive inhibitor molecules. For example, if the ratio of competitive inhibitor molecules to substrate molecules is 1/9, there is a 10% chance of inhibitor molecules binding to the active site. If the ratio is 1/99, there is only a 1% chance of inhibitor binding to the active site. The more substrate present, the less the likelihood of the inhibitor binding to the active site.

6. Measure the amount of product formed in a certain time period, at specific time intervals. The slope of the straight-line portion of the curve near time = zero gives the initial velocity, v_0, as shown in text Figure 5.2.

7.

$$v_0 = \frac{V_{max}[S]}{K_m + [S]};$$ Solving for V_{max} gives:

$$V_{max} = \frac{v_0(K_m + [S])}{[S]}.$$ Using the data given:

$$V_{max} = \frac{0.16 \text{ μmol/min } (2.0 \times 10^{-5}\text{M} + 0.15 \text{ M})}{0.15 \text{ M}}$$

$$V_{max} = \frac{0.024 \text{ μmol/min M}}{0.15 \text{ M}} = 0.16 \text{ μmol/min}$$

(This V_{max} indicates that the substrate concentration used (0.15 M) must have virtually saturated the enzyme. Thus, for $[S] = 2.0 \times 10^{-4}$,

$$v_0 = \frac{0.16 \text{ μmol/min } (2.0 \times 10^{-4}\text{M})}{2.0 \times 10^{-5} \text{ M} + 2.0 \times 10^{-4}\text{M}}$$

$$v_0 = 0.145 \text{ or } 0.15 \text{ μmol/min}.$$

8. A synthase is not the same as a synthetase. A synthase is a lyase that catalyzes an addition reaction. The enzyme in the citric acid cycle that catalyzes the addition of an acetyl group from acetyl coenzyme A to oxaloacetate to form coenzyme A and citrate is citrate synthase. A synthetase is a ligase, which requires energy from ATP or an equivalent high-energy molecule. Also in the citric acid cycle is the enzyme succinyl CoA synthetase. It catalyzes the formation of succinyl CoA from succinate and coenzyme A. GTP is required for energy. In the cycle, however, the reverse of this reaction occurs.

9. Such concentrations allow the enzyme to respond proportionally to changes in substrate concentrations in the cell and prevent the accumulation of substrate under changing metabolic conditions.

10. (a) From the velocity data given, the V_{max} appears to be 50.0 μmol/min. (b) Use that value, and calculate K_m from the data determined from any single measurement. For example, use measurement number 3, for which $[S] = 1.0 \times 10^{-4}$ and $v_0 = 41.0$ μmol/min. This data will also be used in part (c).

$$v_0 = \frac{V_{max}[S]}{K_m + [S]};$$ Solve for K_m:

$$K_m = \frac{V_{max}[S]}{v_0} - [S];$$ Use V_{max} and the data from measurement number 3:

$$K_m = \frac{(50.0 \text{ μmol/min}) (1.0 \times 10^{-4}\text{M})}{41.0 \text{ μmol/min}} - 1.0 \times 10^{-4} \text{ M}$$

$$K_m = 2.2 \times 10^{-5} \text{ M}.$$

(c) A 30.0% decrease in the velocity (41.0 μmol/min) corresponds to a velocity of 28.7 μmol/min for $[S] = 1.0 \times 10^{-4}$ M. This gives an apparent K_m (K_m^{app}) of:

$$K_m^{app} = \frac{(50.0 \text{ μmol/min}) (1.0 \times 10^{-4}\text{M})}{28.7 \text{ μmol/min}} - 1.0 \times 10^{-4} \text{ M}$$

$$K_m^{app} = 7.42 \times 10^{-5} \text{ M}.$$

Use the relationship $K_m^{app} = K_m(1 + [I]/K_i)$:

$$K_i = \frac{K_m[I]}{K_m^{app} - K_m}.$$ Substitute the data from measurement number 3:

$$K_i = \frac{(2.2 \times 10^{-5} \text{ M}) (1.4 \times 10^{-4}\text{M})}{(7.42 \times 10^{-5} \text{ M}) - (2.2 \times 10^{-5}\text{M})};$$ $K_i = 5.9 \times 10^{-5} \text{ M}.$

D. Additional Problems

1. The rate of 30. mmoles/12 min = 2.5 mmoles of product C formation per minute. An International Unit (I. U.) is defined as the number of micromoles of substrate transformed per minute. Therefore, the number of I. U. is:

$$I.U. = 2.5 \ \frac{\text{mmol C}}{\text{min}} \times \frac{10^3 \mu\text{mol C}}{\text{mmol C}} \times \frac{2 \ \mu\text{mol A}}{1 \ \mu\text{mol C}}$$

$$I.U. = 5.0 \times 10^3 \mu\text{mol A}/\text{min} = 5000 \ \text{I.U.}$$

2. The specific activity of an enzyme solution is the number of International Units per mg of protein in the solution used.

Specific Activity = 5000 I. U./2.5×10^{-4} mg protein;

Specific Activity = 2.0×10^7 µmol A transformed per minute per milligram of protein in the solution.

3. The catalytic constant, k_{cat}, or turnover number is the number of micromoles of substrate transformed per second per micromole of active sites. To determine the catalytic constant, convert the specific activity to a per second-basis and determine the number of micromoles of enzyme as shown:

$$k_{cat} = \frac{(5.0 \times 10^3 \mu\text{mol A}/\text{min}) \times 1 \min /60 \sec}{2.5 \times 10^{-4} \text{ mg E} \times \dfrac{1 \text{ mmol E}}{12{,}500 \text{ mg E}} \times \dfrac{10^3 \mu\text{mol E}}{1 \text{ mmol E}}};$$

$$k_{cat} = 4.2 \times 10^6 \sec^{-1}.$$

Note that you can not cancel micromoles of A by micromoles of S. The unit $\sec^{-1}$ actually means micromoles of substrate transformed per second per micromole of active sites. Remember that the enzyme in this problem was reported to have just one active site per molecule.

4. Since the data do not reveal V_{max}, determine it and K_m using a Lineweaver-Burk plot. First, put the kinetic data in reciprocal form as shown:

Measurement Number	1/[S], M^{-1}	$1/v_0$, min/µmol
1	5.88×10^5	0.100
2	2.63×10^5	0.050
3	8.33×10^4	0.022
4	4.35×10^4	0.017
5	1.18×10^4	0.012

Figure 5.1 shows the plot of this data. From the graph, the y-intercept is 0.010 min/µmol. Since the y-intercept = $1/V_{max}$,

$$V_{max} = 1/0.010 \text{ min/µmol} = 100 \text{ µmol/min}.$$

The y-intercept is -6.7×10^4 M^{-1}. Therefore, -6.7×10^4 $M^{-1} = -1/K_m$;

$$K_m = 1/6.7 \times 10^4 \text{ M}^{-1} = 1.5 \times 10^{-5} \text{ M}.$$

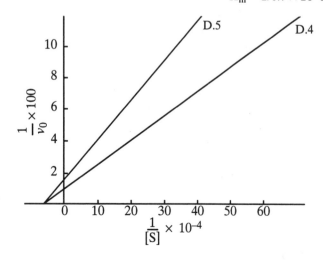

Figure 5.1 Lineweaver-Burke Plot for Problems D.4 and D.5.

5. Put the data in reciprocal form and plot it on the same graph used in the previous problem. This gives a straight line with the same x-intercept (approximately -6.7×10^4 M^{-1}) as shown in Figure 5.1. This indicates noncompetitive inhibition. The slope of this line is greater than that of the uninhibited line. You may obtain values which are slightly different due to differences in graphing techniques. You should also be aware that noncompetitive inhibition is very rare. The data in this problem were arbitrarily chosen and are not from an actual experiment.

CHAPTER 6

A. True–False

1. <u>False.</u> It is a nucleophile. Having unpaired electrons, it is attracted to an electron-deficient atom. It is named for that to which it is attracted. It is a nucleus lover.

2. <u>True.</u> The carbon atom carries a positive charge. It is also called a carbonium ion.

3. <u>True.</u>

4. <u>False.</u> Active site residues usually include several hydrophilic side chains. Such residues compose the catalytic center of the enzyme.

5. <u>True.</u> Covalent bonds form between a reactant and some enzyme active site group. One product is then released from the enzyme prior to release of the other product.

6. <u>False.</u> Moderately weak binding of reactants is necessary for efficient catalysis.

7. <u>False.</u> Entropy decreases as reactants are collected and oriented in a particular way, losing freedom of motion.

8. <u>True.</u> Such analogs are potent enzyme inhibitors.

9. <u>True.</u> Activation of chymotrypsinogen involves removal of two dipeptide segments of the zymogen chain leaving three peptide chains held together by a total of five disulfide bonds.

10. <u>True.</u> The catalytic triad consists of Asp, His, and Ser.

B. Short Answers

1. mechanism

2. transition state

3. activation energy, E_a

4. diffusion-controlled

5. proximity effect

6. thermodynamic pit

7. low-barrier hydrogen bonds

8. serine

9. Transition state stabilization

10. acid-base catalysis; covalent catalysis; proximity effects; transition state stabilization

C. Problems

1. (a) The carbonyl carbon atom has partial positive character and, therefore, would be attacked by the nucleophilic Y:$^{\ominus}$ species.

   (b) The leaving group would be an anion. Extra electrons are brought to the carbon atom by the Y:$^{\ominus}$ nucleophile, so electrons must leave (eg. as X:$^{\ominus}$) to achieve a charge balance.

2. Intermediates are metastable species with lifetimes long enough (10^{-14}–10^{-13} sec, or longer) to allow them to be detected by sophisticated techniques. Transition states have even shorter lifetimes and have yet to be detected. The formation of an intermediate will involve a second activation energy barrier to be overcome before product is formed.

3. (a) Ala + Cys—Lys—Met—Phe—Arg—Ala + Tyr—Gly

   (b) Ala—Cys—Lys + Met—Phe—Arg + Ala—Tyr—Gly

   (c) Ala—Cys—Lys—Met—Phe + Arg—Ala—Tyr + Gly

   (d) Ala—Cys—Lys—N—CH—C=O + Phe—Arg—Ala—Tyr—Gly

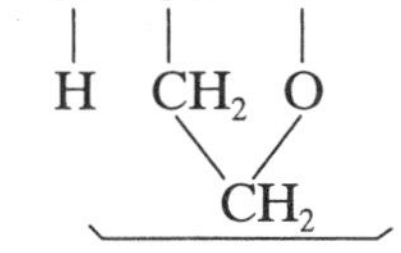

   homoserine lactone

4. An enzyme lowers the overall activation energy by providing a multi-step pathway in which the steps have lower activation energies than those of corresponding stages in the nonenzymic reaction.

5. In addition to the Asp-His-Ser catalytic triad, each enzyme has a binding site or specificity pocket which fits the side chain of the residue recognized. See text Figure 6.23.

6. Enteropeptidase is a duodenal enzyme that catalyzes removal of a N-terminal hexapeptide from trypsinogen to create active trypsin. Trypsin then catalyzes the activation of other zymogens in the intestinal tract.

7. The bond cleaved that converts trypsinogen into trypsin is the Lys 6 - Ile 7 peptide bond. Since trypsin catalyzes the cleavage of peptide bonds in which the carbonyl group is donated by lysine or arginine, the Lys 6 residue would be recognized by trypsin.

8. Although trypsin inhibitor contains residues that trypsin recognizes, the fit between the inhibitor and trypsin is very tight due partly to a complex network of hydrogen bonds. The fit is so tight that water can not enter the active site to participate in hydrolysis.

9. This product was formed when DIFP reacted with an active site serine residue. Because the serine is involved in the catalytic mechanism, the enzyme was inactivated when the serine side chain was no longer available.

10. (a) The substrate binds in the active site with the carbonyl carbon of the scissile peptide bond close to the oxygen of the serine in the catalytic triad (Ser-195).

    (b) A proton is removed from the -OH of Ser-195 by attraction of the >N: of the His-57 side chain. This creates a more nucleophilic oxygen on Ser-195 which then attacks the carbonyl carbon of the scissile peptide bond.

    (c) The carbonyl oxygen of the peptide bond becomes an oxyanion and moves into the oxyanion hole and is hydrogen bonded to -N-H groups of peptide bonds of Gly-193 and Ser-195.

    (d) The proton abstracted by His-57 is donated to the nitrogen of the peptide bond, which causes its cleavage and the release of the first product, a free amine.

    (e) Water enters the active site and donates a proton to His-57, resulting in formation of $OH^{\ominus}$, which performs a nucleophilic attack on the carbonyl carbon of the serine ester group.

    (f) His-57 donates the proton obtained from water to the Ser-195 oxygen, thus releasing the second product, a carboxylate ion, and regenerating the enzyme.

D. Additional Problems

1. As you have learned, the maximum rate of reaction for an enzyme-catalyzed system occurs when excess substrate (saturating levels) is present. With excess substrate present, the rate does not change and zero order kinetic conditions with respect to substrate concentration are observed. In this case, the rate = $k[S]^0 = k$. Thus, the rates can be used in the Arrhenius equation.

$$\ln k = -E_a/R \times 1/T + \ln A$$

   Write two equations using k_1 and T_1 and k_2 and T_2, respectively. Subtract the first equation from the second and solve for E_a.

$$\ln k_2 - \ln k_1 = E_a/R \times (1/T_1) - E_a/R \times (1/T_2).$$

   Rearrange this to

$$\ln (k_2/k_1) = E_a/R(1/T_1 - 1/T_2), \text{ and solve for } E_a.$$

$$E_a = \frac{R \times 1n \; k_2 / k_1}{\dfrac{1}{T_1} \; - \; \dfrac{1}{T_2}} = \frac{(8.314 \text{ J/mol K}) \, (2.303 \log 36.6/25.2)}{(1/293 - 1/303) \text{ K}^{-1}} =$$

$$= 2.7 \times 10^4 \text{ J/mol K or about 27 kJ/mol.}$$

2. The pH profile shown in Figure 6.1 indicates an acid-base mechanism involving the two aspartate residues in the active site. One aspartate has a pK_a of about 1.4, is therefore unprotonated at stomach pH, and thus acts as a proton acceptor (a general base). The other aspartate has a pK_a of about 4.3, is protonated at stomach pH, and acts as a proton donor (a general acid). Optimal activity occurs at the intermediate pH of about 2.7. The side chain of free aspartic acid has a pK_a of 3.9, so the pK_a values of both aspartate side chains are substantially altered by neighboring groups in the enzyme's active site.

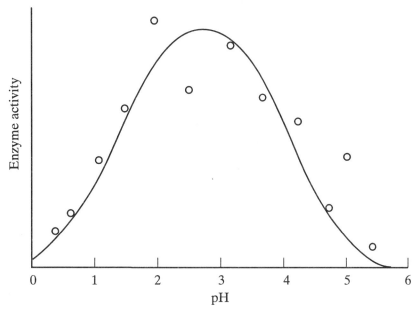

Figure 6.1 pH Profile of Pepsin.

3. Figure 6.2 indicates that one product, P, appears before the other, indicating a multi-step mechanism. This burst of the first product suggests that the substrate is cleaved with one product released rapidly. The remaining part of the substrate is likely covalently bound to a group in the active site of the enzyme and its release is controlled by the rate-limiting step of the mechanism.

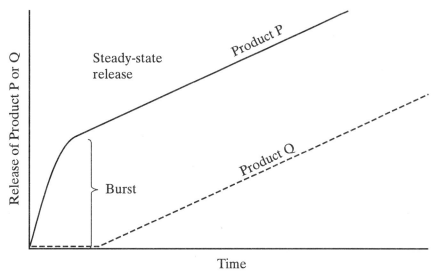

Figure 6.2 Burst Kinetics.

4. For the non-enzymatic bimolecular reaction,

$$\text{rate} = k_2[A][B]; \qquad k_2 = \frac{\text{rate}}{[A][B]}$$

from the data given,

$$k_2 = \frac{2.5 \times 10^{-2}\,\text{mmol}/2\,\text{min}}{[0.0100][0.0100]\,\text{M}^2}$$

$$k_2 = 1.25 \times 10^2\,\text{mmol}/(\text{min}\ \text{M}^2)$$

The enzyme-catalyzed reaction is unimolecular because reactants and catalytic groups are positioned near each other in the active site. Therefore,

$$k_1 = \text{rate}/[A][B] = \frac{4.0\,\text{mmol}/2\,\text{min}}{[0.0100]\,\text{M}}$$

$$k_1 = 2.0 \times 10^2\,\text{mmol}/(\text{min}\ \text{M})$$

Since the effective molarity is the ratio of k_1/k_2,

$$\text{effective molarity} = \frac{2.0 \times 10^2\,\text{mmol}\ \text{min}^{-1}\ \text{M}^{-1}}{1.25 \times 10^2\,\text{mmol}\ \text{min}^{-1}\ \text{M}^{-2}}$$

Effective molarity $= 1.6 \times 10^4$ M.

Note that the proximity effect is usually not the sole enzymatic catalytic mode.

5. Potential substrates with two, three, and four NAG units (A-B, A-B-C, and A-B-C-D) are rather stable to lysis in the presence of lysozyme. The highest rate of hydrolysis occurs with NAG_6 and the rate does not increase with additional NAG units. The specificity, therefore, is for the oligomer with six units in the active site (A-B-C-D-E-F) and all six sugar-binding sites must be occupied for maximal activity.

Since the A-B-C-D oligomer is not cleaved, none of these bonds (A-B, B-C, or C-D) are candidates. Cleavage of NAG_5 suggests that the D-E bond is cleaved. The data indicate that lysozyme cleaves the bond indicated below in bacterial cell walls.

NAG-NAM-NAG-NAM-NAG-NAM

CHAPTER 7

A. True–False

1. <u>False.</u> The iron of iron-sulfur clusters is not part of a heme group. It is, in fact, referred to as "non-heme iron."

2. <u>True.</u> These are made from metabolites which occur normally in the organism.

3. <u>False.</u> Water-soluble vitamins are readily excreted from the body and do not normally build up to toxic levels. Lipid vitamins (fat-soluble vitamins) can accumulate in the fatty tissues of the body and lead to hypervitaminosis.

4. <u>False.</u> Pellagra is due to a deficiency of niacin required for synthesis of $NAD^{\oplus}$ and $NADP^{\oplus}$ coenzymes.

5. <u>True.</u> The coenzyme form, thiamine pyrophosphate (TPP), is required by certain decarboxylase enzymes and by some transketolases.

6. <u>True.</u> The anionic tail helps the coenzyme bind in the enzyme active site.

7. <u>False.</u> Vitamin K is a fat-soluble vitamin, but it plays a role in proper synthesis of proteins involved in the blood-clotting mechanism. Vitamin E is a lipid vitamin that exhibits antioxidant activity.

8. <u>True.</u> Coenzyme QH_2 is a weaker reducing agent than NADH. So, under standard conditions, the reaction will proceed as:

$$NADH + H^{\oplus} + \text{Coenzyme Q} \rightarrow NAD^{\oplus} + \text{Coenzyme } QH_2$$

9. <u>True.</u> The Fe^{3+} can accept an electron and become Fe^{2+}, which can donate an electron and return again to the Fe^{3+} state.

10. <u>False.</u> Vitamins A and K act as coenzymes, but D and E do not. Vitamin D is a generic descriptor of steroids that exhibit the biological activity of cholecalciferol, which includes a role in the absorption and deposition of calcium phosphate. Vitamin E is a generic descriptor for compounds that exhibit the biological activity of α-tocopherol, which includes antioxidant activity and a role in normal growth and fertility.

B. Short Answers

1. coenzymes; essential inorganic ions

2. apoenzyme

3. holoenzyme

4. reagents with chemical properties unlike those of any of the amino acid side chains in the enzyme active site

5. reactive center

6. cosubstrate

7. prosthetic group

8. ylid

9. water-soluble; lipid-soluble

10. protein coenzymes; metal ions; iron-sulfur clusters; heme groups

C. Problems

1. Metal-activated enzymes are those that require metal ions such as K^+, Mg^{2+}, or Ca^{2+} for their activity. These ions may aid the binding of substrates to the enzyme. Metalloenzymes contain firmly-bound ions, such as those of iron, zinc, copper and cobalt, in their active sites . These ions do not dissociate from the enzyme and usually participate in the reaction catalyzed.

2. Both types of coenzymes, cosubstrates and prosthetic groups, provide the active sites of enzymes with reactive groups not supplied by amino acid side chains in the active site of the enzyme.

3. *Coenzyme A* acts as a cosubstrate involving the transfer of acyl groups.

   NAD acts as a cosubstrate in oxidation-reduction reactions involving dehydrogenase enzymes.

   The vitamin biotin is a prosthetic group for ATP-dependent enzymes that catalyze carboxylation reactions.

   Thiamine pyrophosphate (TPP) is derived from vitamin B_1 and is a prosthetic group for enzymes that catalyze decarboxylation or the transfer of two-carbon units containing a carbonyl group.

   Tetrahydrofolate is derived from the vitamin folate and serves as cosubstrate for enzymes that catalyze the transfer of one-carbon groups including formyl, hydroxymethyl, and methyl groups.

   ATP is a metabolite coenzyme that acts as a cosubstrate for enzymes that catalyze the transfer primarily of phosphoryl or nucleotidyl groups.

   FAD is derived from vitamin B_2, riboflavin, and serves as a cosubstrate for dehydrogenase enzymes that catalyze certain oxidation-reduction reactions.

4. Vitamin C exists as a lactone, which is an internal (cyclic) ester that forms when a carboxyl group reacts with a hydroxyl group on the same molecule. A molecule of water is eliminated and an ester, the lactone, results. Vitamin C is a required reducing agent in the hydroxylation of collagen.

5. The mechanism by which NAD^+ becomes reduced is by the acceptance of a hydride ion, H^-, which neutralizes the positive charge of NAD^+ and forms NADH. A proton is also removed from the substrate, but is released into solution. So, NADH and H^+ are products formed as the cosubstrate NAD^+ is reduced and the substrate is oxidized. Note that this is a two-electron oxidation-reduction.

6. It is an ordered sequential mechanism, not a ping-pong mechanism. Cosubstrate NAD^+ binds to the enzyme first, producing the holoenzyme, which then binds the substrate lactate. The product pyruvate is released first, followed by the release of NADH. See text figure 7.10.

7. The substrate lactate is properly positioned by the formation of an ion pair with the positively-charged side chain of Arg-171 and its own negatively-charged carboxylate group. With lactate positioned, a proton is abstracted from the hydroxyl group on C-2 of lactate by the nearby side chain of his-195. As the electron pair that held the hydrogen of the hydroxyl group shifts back to the C-2 carbon of lactate, the hydrogen atom also on C-2 can leave with its electrons as a hydride ion and add to C-4 of the pyridine ring of NAD^+. Oxidation of the substrate is then complete and the product pyruvate is released. Note these details in text Figure 7.9.

8. FAD and FMN are held more tightly by their apoenzymes than is $NAD^{\oplus}$ by its apoenzymes. $FADH_2$ and $FMNH_2$ are rapidly oxidized in solutions containing oxygen, but are protected from this oxidation by remaining tightly bound in the active site of an enzyme. NADH does not have to be protected in this way since it is not susceptible to this kind of oxidation. Figure 7.1 shows the mechanisms by which $NAD^{\oplus}$ and FAD are reduced. Note that FAD is reduced by two one-electron transfers.

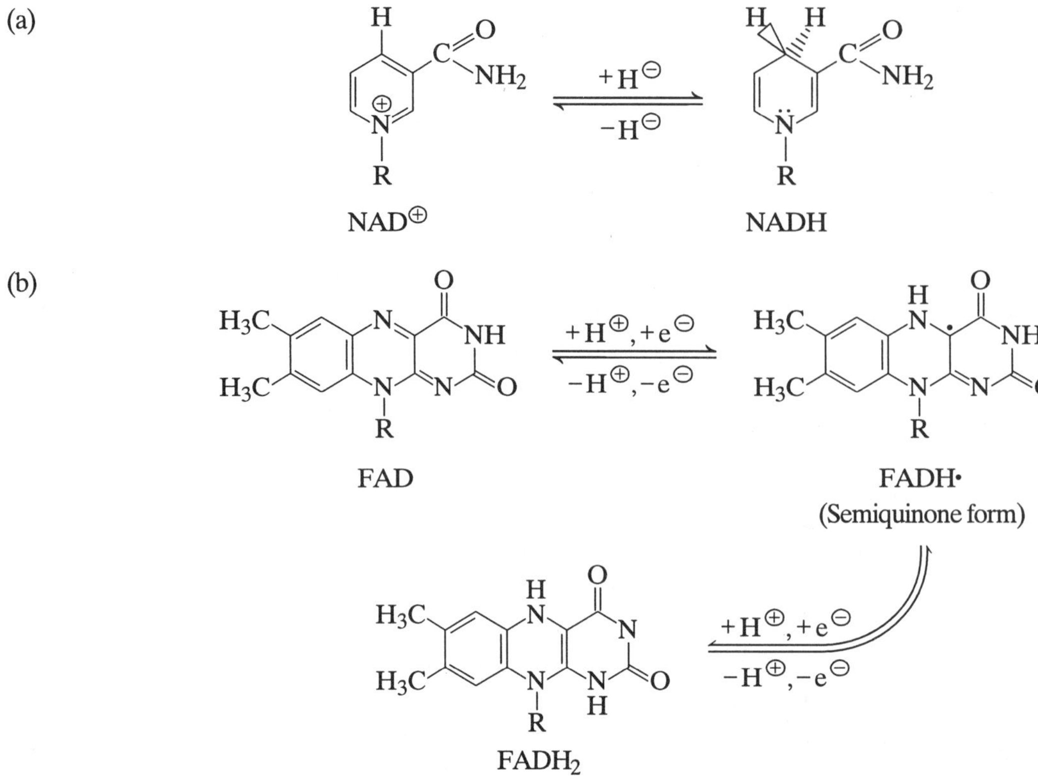

(a)

(b)

Figure 7.1 Mechanisms of Reduction of $NAD^{\oplus}$ and FAD.

9. Coenzyme Q (ubiquinone or Q) has a tail of six to 10 isoprenoid units that makes the coenzyme soluble in lipid membranes. Q is located in the inner mitochondrial membrane, where it participates in electron transport.

10. The reduction potential is a measure of the tendency to receive an electron from another substance. The cytochromes, even within a given class, have different reduction potentials due to different chemical environments (different amino-acid side chains) around the heme groups.

D. Additional Problems

1. Biotin becomes bound in the enzyme active site by forming an amide linkage with its carboxyl group and the ε-amino group of lysine. This forms the prosthetic group biocytin. Some texts indicate that ATP may react with bicarbonate to form carbonylphosphate, an active form of CO_2 that reacts with enzyme-bound biotin to form N-carboxybiotin. An active site base abstracts a proton from the β-carbon of pyruvate, forming a carbanion that performs a nucleophilic attack on N-carboxybiotin as indicated in Figure 7.2. Oxaloacetate is formed and the enzyme is regenerated.

2. Cancer involves uncontrolled cell division, which requires DNA synthesis. DNA is composed of the nucleotides dAMP, dGMP, dTMP, and dCMP. The synthesis of dTMP, from dUMP, is catalyzed by the enzyme thymidylate synthase, which requires methylene tetrahydrofolate as the methyl group donor. In the reaction, dihydrofolate is produced and must be regenerated to tetrahydrofolate by the action of dihydrofolate reductase. If this enzyme is inhibited, supplies of tetrahydrofolate will be decreased, dTMP will decrease, and DNA synthesis and cell division will not occur. The antifolate drug methotrexate, an inhibitor of dihydrofolate reductase, has been used extensively in the treatment of cancer.

3. The cofactor that would be affected is biotin. Raw egg whites contain the protein avidin, which has a strong affinity for biotin and prevents absorption of biotin from the intestinal tract. Without sufficient biotin, reactions such as the production of oxaloacetate, required for initiation of the Krebs cycle, would not be possible and energy production would be impaired.

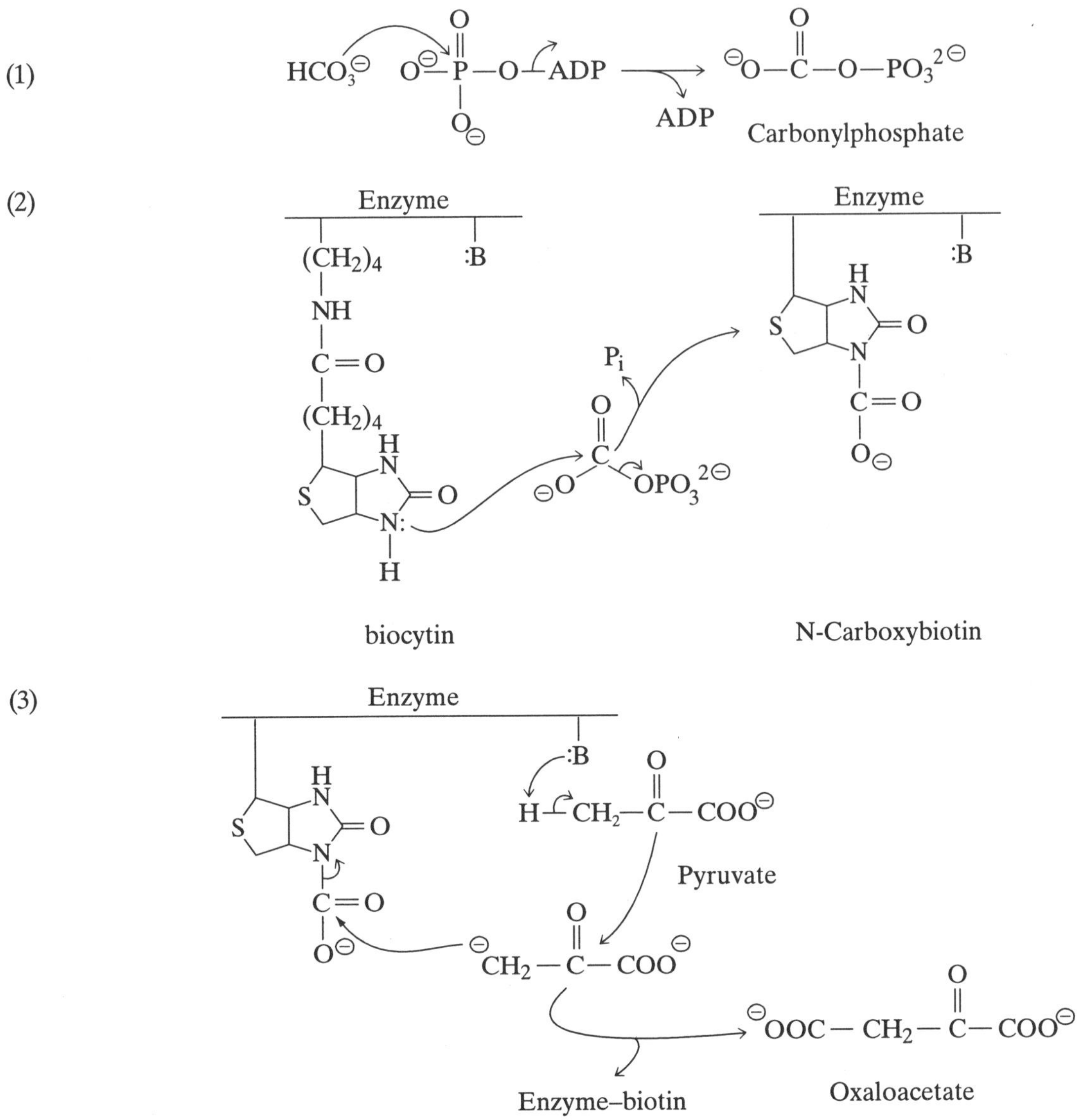

(1)

(2)

biocytin · N-Carboxybiotin

(3)

Enzyme–biotin · Oxaloacetate

Figure 7.2 Mechanism of Pyruvate Carboxylase.

4. (a) Pyridoxal phosphate (PLP) is the coenzyme that forms Schiff bases during catalysis by enzymes such as transaminases. (b) An aldimine is a Schiff base formed from the condensation of a primary amine and an aldehyde. (c) An internal aldimine is formed by interaction of the aldehyde group of PLP with the ε-amino group of lysine in the active site of the enzyme as shown in text Figure 7.17. The external aldimine is formed by the interaction of the α-amino group of the substrate amino acid with the internal aldimine. The net result is a displacement of the active site lysine by the substrate amino acid. After hydrolysis of the intermediate, an α-keto acid product is formed as well as pyridoxamine phosphate (PMP). This form of the coenzyme serves as an amine group donor to an α-keto acid substrate to complete the transamination reaction.

5. (a) The disease may develop due to a deficiency of vitamin B_{12}. (b) The coenzyme forms are methylcobalamin, which participates in transfer of methyl groups, and 5′-deoxyadenosylcobalamin, which participates in intramolecular rearrangements. (c) The vitamin contains cobalt, and a heme-like group called corrin. Humans require very small amounts (about 3 micrograms per day). To be absorbed, the vitamin must combine with a protein called intrinsic factor that is secreted in the stomach. In older people, sufficient absorption of vitamin B_{12} may not occur due to inadequate secretion of the intrinsic factor, even though there is an adequate supply of the vitamin in their diet. Since vitamin B_{12} is not made by plants, strict vegetarians may eventually become deficient in this vitamin because they are not eating meat products. Since the liver typically stores an adequate (five to six-year) supply of vitamin B_{12}, onset of the disease is not immediate.

CHAPTER 8

A. True–False

1. <u>False.</u> It is a ketotriose.

2. <u>False.</u> Unlike the amino acids, most natural sugars have the D-stereochemistry like that of D-glyceraldehyde.

3. <u>True.</u> Note that the configuration of the anomeric carbon of cyclic structures is not fixed, but changes in an equilibrium reaction. The configurations of the non-anomeric chiral carbons are fixed.

4. <u>True.</u> Those that form 5-member rings are the furanoses.

5. <u>True.</u> All three are homoglycans of glucose.

6. <u>False.</u> Amylopectin is the branched polymer of starch.

7. <u>True.</u> So is cellobiose.

8. <u>False.</u> All but sucrose are reducing sugars.

9. <u>False.</u> It has α-(1 → 4) linkages like starch but glycogen is more highly branched than amylopectin.

10. <u>True.</u>

B. Short Answers

1. oligosaccharide—specifically, a pentasaccharide

2. anomeric carbon

3. anomers

4. aglycone

5. limit dextrin

6. glycoforms

7. chitin; N-acetyl-D-glucosamine

8. peptidoglycan

9. mucins

10. D-psicose (For other correct answers, see text Figure 8.5.)

C. Problems

1.

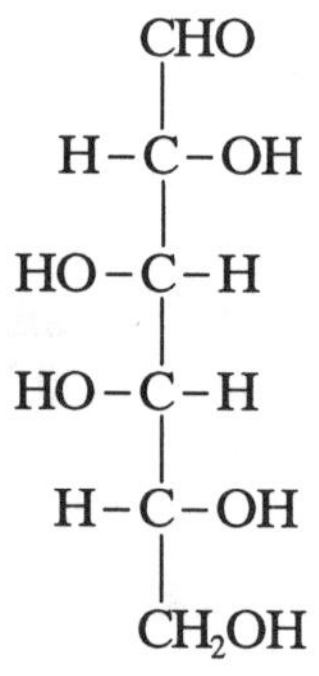

D-galactose

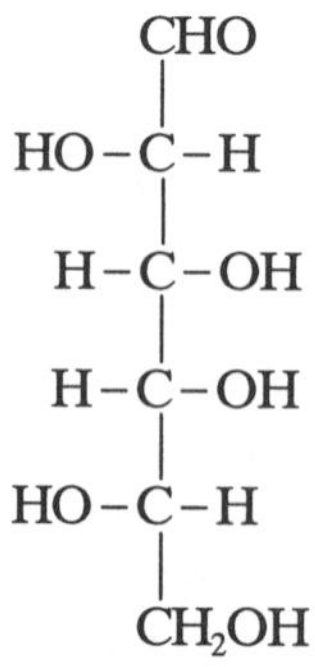

L-galactose (the enantiomer)

2.

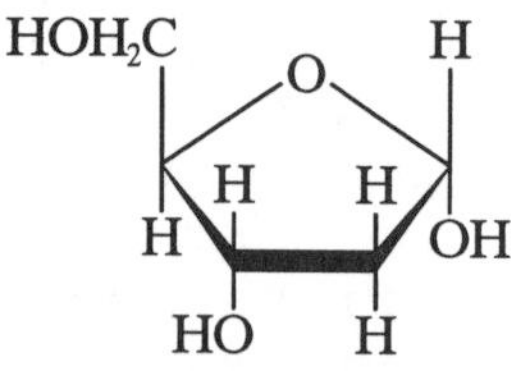

2-deoxy-α-D-ribofuranose

3. If 4% of the glucose residues are involved in branches, then 212 residues $\times$ 0.04 residues/branch = 8 branches. They occur about every 25 glucosyl residues.

4. The "head to head" linkage of glucose to fructose in the sucrose molecule ties up both anomeric carbons. Consequently, the anomeric carbons can not open up to give the open chain form required for the reducing sugar reactions.

5.
   a. D-glucose

   b. *N*-acetyl-D-glucosamine

   c. D-galactose and D-glucose

   d. D-glucose

   e. D-glucose

6.

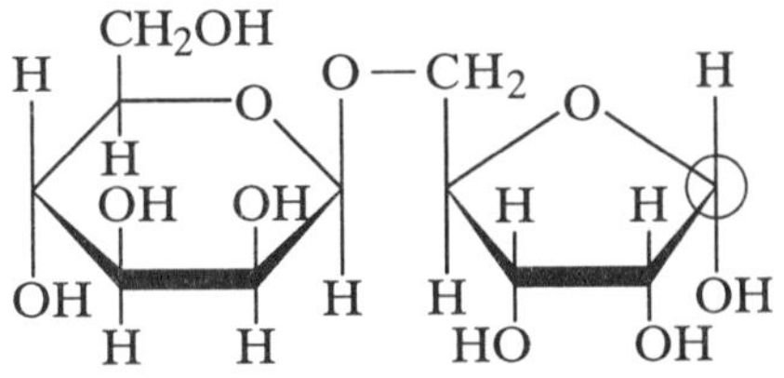

Yes. The ribose anomeric carbon (circled) is free to undergo ring opening.

7.

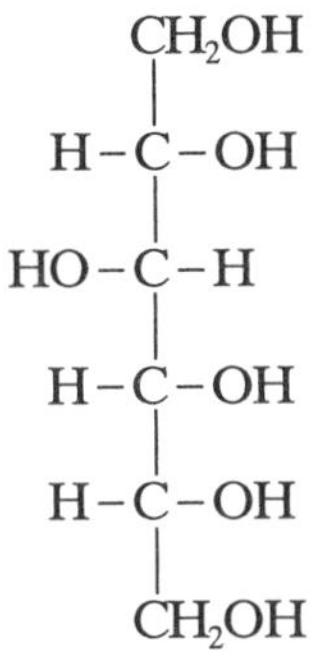

D-glucitol 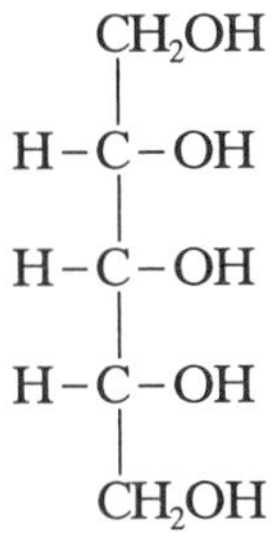   D-ribitol

They can not form cyclic structures since there is no aldehyde or ketone group available.

8. (a) The molar mass of glucose, $C_6H_{12}O_6$, is 180 g, but each internal glucosyl residue is less than 180 by the mass of a water molecule, eliminated when polymerization occurred. (This situation is just like that involving the mass of an amino acid residue in a protein chain.) Therefore, the residue mass of glucose is 180 – 18 = 162 g/mole. If the molar mass of glycogen is about 3.0×10^6 g, then the number of glucosyl residues is:

$$3.0 \times 10^6 \text{ g} \times 1 \text{ residue}/162 \text{ g} = 1.9 \times 10^4 \text{ residues.}$$

(b) Your text indicates that branch points occur in glycogen every 8 to 12 residues (an average of about 10%). Therefore, the number of glucosyl residues at branch points is:

$$1.9 \times 10^4 \text{ residues} \times 1 \text{ branch}/10 \text{ residues} = 1900 \text{ residues.}$$

9. The moles of glycogen in one pound:

$$\frac{454 \text{ g}}{\text{lb}} \times 1 \text{ mole}/3.0 \times 10^6 \text{ g} = 1.5 \times 10^{-4} \frac{\text{mole}}{\text{lb}}.$$

The molecules of glycogen will be:

$$1.5 \times 10^{-4} \text{ mole} \times 6.02 \times 10^{23} \text{ molecules/mole} = 9.0 \times 10^{19} \text{ molecules.}$$

10. (a) There are 4 chiral centers. (b) Therefore, there should be 2^4 or 16 stereoisomers. (c) Text Figure 8.3 shows only 8 aldohexoses, but there are an equal number of the L- series aldohexoses (the enantiomers of the 8 D-aldohexoses).

11. There are five chiral centers in the Haworth or cyclic structures of the aldohexoses. Therefore, there can be 2^5 or 32 stereoisomers. The extra forms are due to the two anomeric forms (α and β) possible for each of the 16 stereoisomers of the previous problem.

D. Additional Problems

1. Your text indicates that, in aqueous solution, 36% of the glucose molecules are in the α-anomeric form and nearly all the rest, 64% consists of the β-anomeric form. Using these quantities and the specific rotation values of each, calculate the equilibrium optical rotation that will eventually be reached in the aqueous solution due to mutarotation.

$$112.2°(0.36) + 18.7°(0.64) = 52°$$

(This neglects the tiny amount (less than 0.1%) of open chain form that might be present.)

2. Since $[\alpha] = A/(l)(c)$, rearrangement to solve for c gives:

$$c = A/[\alpha](l) = \frac{46.8°}{(66.5°)(2.5)} = 0.28 \text{ g/mL}$$

3. (a) The length of this segment of amylose helix is:

$$(900 \text{ residues})(1 \text{ turn}/6 \text{ residues})(0.8 \text{ nm/turn}) = 120 \text{ nm}.$$

(b) Amylose forms a left-handed helix that is hydrated inside as well as outside, whereas the α-helix of proteins is right-handed and contains no water molecules in its interior. The α-helix of proteins contains fewer residues (3.6) per turn than does the amylose helix and has more turns in a given linear distance. If it were possible to have a polypeptide chain of 900 amino acid residues all in an α-helix, the length would be:

$$0.15 \text{ nm/residue} \times 900 \text{ residues} = 135 \text{ nm}.$$

4. (a)

D-galactose   D-galactonate

(Note: The two silver diammine complexes contain four ammonia molecules that are released when silver ion is reduced to metallic silver. Two water molecules are formed when two protons react with two hydroxide ions. One of the protons is released from the aldehyde group of the sugar after nucleophilic attack by hydroxide ion on the carbonyl carbon. The other proton comes from the carboxylic acid group formed in this reaction.

(b) The open chain form is in equilibrium with the cyclic anomeric forms. As some open chain form is irreversibly consumed, more will form as a new equilibrium is established. Thus, eventually, all of the sugar will be oxidized.

5. (a) Figure 8.1 shows the formation of a Schiff base between the primary amino group of a protein and the aldehyde group of a sugar (D-glucose, in this example). Formation of the Amadori product occurs as the result of two tautomerizations involving the Schiff base.

(b) Figure 8.2 shows the formation of the cross link between proteins resulting from the reaction of an Amadori product, a glucose molecule, and the primary amino group of a protein. Seven molecules of water are eliminated in this multistep condensation.

6. There are 20 possible disaccharides.

(a) There are four possible disaccharides involving linkages of the anomeric carbons of both sugars. They are: αG and βF (sucrose), αG and αF, βG and αF, and βG and βF.

(b) There are four possible disaccharides involving glucose in the alpha orientation and hydroxyl groups 1, 3, 4, and 6 of fructose.

(c) There are an additional four possible disaccharides involving glucose in the beta orientation and hydroxyl groups 1, 3, 4, and 6 of fructose.

(d) There are four possible disaccharides involving fructose in the alpha orientation and hydroxyl groups 2, 3, 4, and 6 of glucose.

(e) There are four possible disaccharides involving fructose in the beta orientation and hydroxyl groups 2, 3, 4, and 6 of glucose.

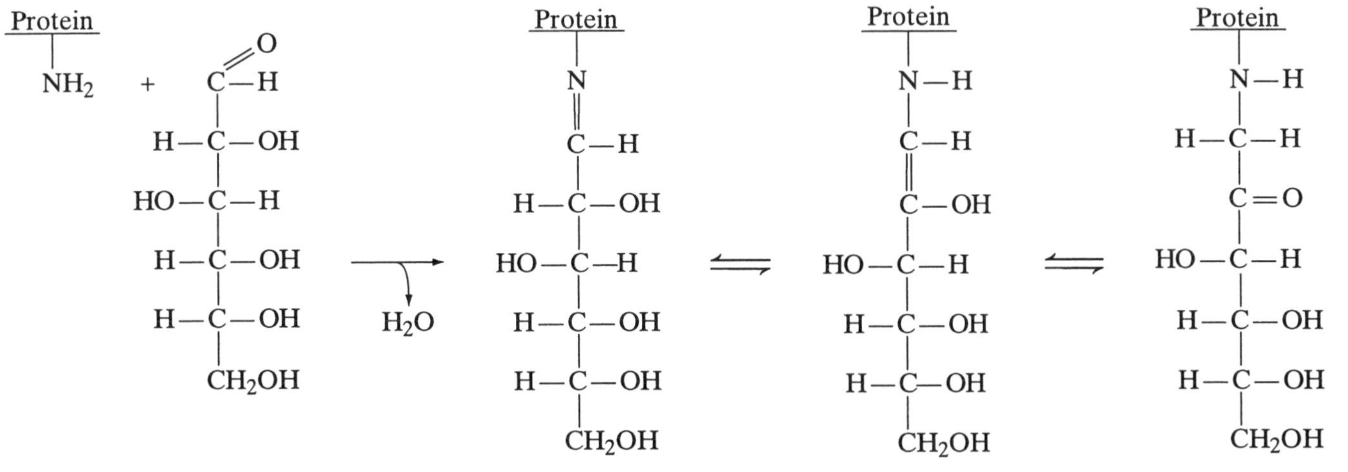

D-glucose Schiff base (Enol form) Amadori product (Keto form)

Figure 8.1 Formation of an Amadori Product from a Sugar and a Protein.

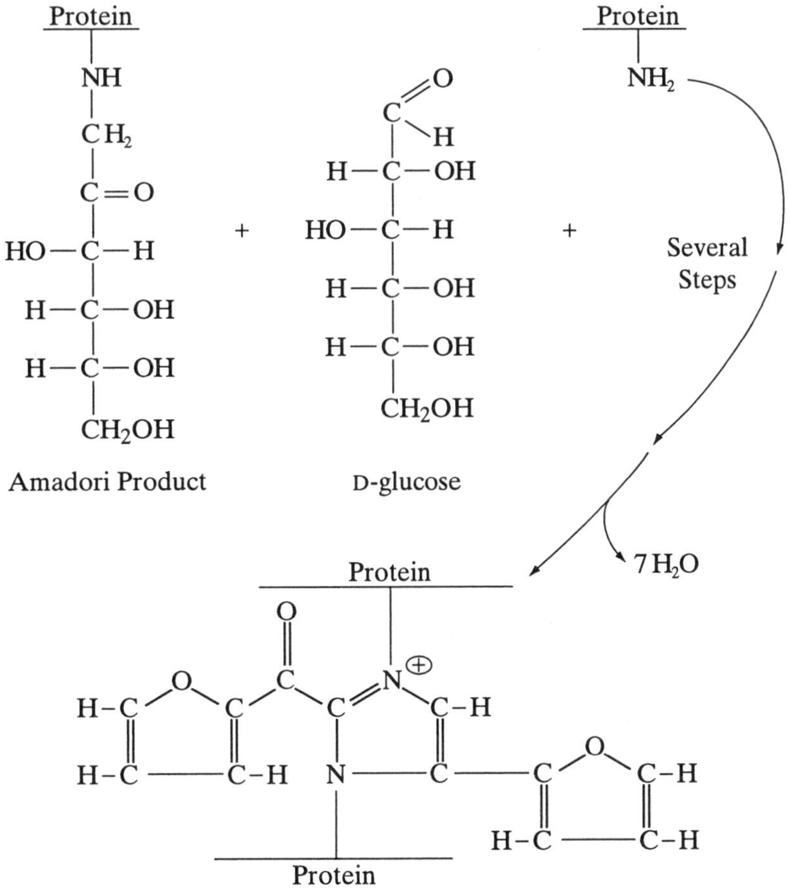

Figure 8.2 Formation of Cross-linked Proteins.

CHAPTER 9

A. True–False

1. <u>False.</u>   Nucleotides contain three different kinds of components: a sugar (ribose or deoxyribose), a nitrogenous base (a purine or pyrimidine), and one or more phosphate groups.

2. <u>True.</u>   The pentose D-ribose is derived from RNA and 2-deoxy-D-ribose is derived from DNA.

3. <u>False.</u>   Uracil and thymine are pyrimidines, but uracil is a component of RNA, but not DNA.

4. <u>True.</u>   It is, in fact, 3′-deoxyadenosine.

5. <u>True.</u>   Different kinases are involved in maintaining the appropriate levels of the different nucleotides and other phosphorus-containing metabolites in the cell.

6. <u>False.</u>   An induced fit between enzyme and substrate occurs when binding of a substrate causes a conformational change in the enzyme active site, giving precise fit to the substrate(s) (or to the transition state for the reaction.) This occurs when adenylate kinase binds its substrates AMP and ATP.

7. <u>True.</u>   Phosphatases catalyze hydrolysis of phosphate esters.

8. <u>True.</u>   cAMP and cGMP are examples of signal nucleotides.

9. <u>True.</u>

10. <u>False.</u>   All the purine and pyrimidine nucleoside diphosphates are accepted as substrates. ATP is usually the phosphoryl group donor under cellular conditions.

B. Short Answers

1. uracil

2. cytosine

3. lactam; lactim

4. dTTP

5. guanylate kinase

6. UDP-glucose

7. ADP-glucose

8. cyclic ADP-ribose

9. Second messengers

10. deoxynucleoside triphosphates (dTTP, dATP, dGTP, dCTP). The corresponding monophosphate residues are actually incorporated in the DNA chain.

C. Problems

1. The structure of cytosine arabinoside is very much like that of deoxycytidine, whose triphosphate is normally required for DNA synthesis. Inside the cell, cytosine arabinoside becomes phosphorylated and is incorporated into a growing DNA chain as araCMP. It is a substrate for cell DNA polymerases, but incorporation of two or more successive araCMP residues causes chain termination. The cell DNA polymerases have difficulty adding the next nucleotide after araCMP has been added to the chain. Therefore, cell division is prevented.

2. ATP and other related nucleotide triphosphates are ionized at cellular pH and apparently form complexes with $Mg^{2\oplus}$ ions in cells. Specifically, a $Mg^{2\oplus}$ ion complexes with the second and third phosphate groups of ATP, forming a β,γ complex.

3. Formation of GDP is favored because the concentration of ATP in cells is usually much higher than the concentration of the other substrate, GMP, or of either product, GDP and ADP.

4. The fact that no product is formed unless both substrates are present suggests a sequential mechanism—both substrates must be in place in the active site before catalysis takes place. This also suggests that covalent catalysis does not occur. No group is transferred from one substrate to a group in the active site of the enzyme prior to its transfer

to the other substrate. If this were not true, ADP could transfer a phosphoryl group to the enzyme and a stoichiometric amount of ADP would appear as a product prior to the addition of AMP.

5. The enzyme nucleoside diphosphate kinase catalyzes the formation of nucleoside triphosphates from the corresponding diphosphates with ATP as the phosphoryl group donor.

$$NDP + ATP \longrightarrow NTP + ADP$$

The enzyme accepts all the common ribonucleoside and deoxyribonucleoside diphosphates as substrates.

6. α-D-Glucose 1-phosphate is a hemiacetal phosphate and has a higher glucose-transfer potential (larger negative free energy of hydrolysis) than the glucose 6-phosphate, which is a phosphoester. Mechanistically, it is the 1-phosphate that reacts with the α-phosphorus of UTP to form UDP-glucose.

7. This hydrolysis of PP$_i$ helps drive the formation of UDP-glucose to completion by removal of one of the products of that reaction. The presence of cellular pyrophosphatases assures that PP$_i$ is always broken down to 2 P$_i$, which drives all pyrophosphate cleavages toward completion.

8. Extracellular signals include those supplied by hormones, which are released into the bloodstream in response to stimuli. The hormones then interact with specific receptors on the membranes of target cells. This interaction may trigger the formation of a second messenger such as cAMP, which travels within the cell to a target enzyme and influences its activity.

9. Alarmones are signal nucleotides that accumulate during times of metabolic stress. They influence gene expression and regulate activities of certain enzymes. Guanosine 5'-diphosphate 3'-diphosphate, ppGpp, and guanosine 5'-triphosphate 3'-diphosphate, pppGpp, are alarmones that have been found in bacteria.

10. Cyclic ADP-ribose is a signal nucleotide formed from $NAD^{\oplus}$. It induces the release of $Ca^{2\oplus}$ ions in certain cells.

D. Additional Problems

1. (a) Xanthine (2,6-dioxopurine) is a purine base formed by the oxidative deamination of guanine. (b) The adrenaline-producing action of cyclic AMP ceases when the phosphodiesterase catalyzes the conversion of cyclic AMP to the 5'-AMP. Caffeine acts as an inhibitor of the phosphodiesterase, probably due to its molecular similarity to the purine base adenine of cyclic AMP. This inhibition prolongs the adrenalin-producing effect of cyclic AMP.

2. Thymidylate synthase catalyzes the methylation of deoxyuridine-5'-monophosphate (dUMP) to thymidine-5'-monophosphate (dTMP). The inhibition of this enzyme by 5-fluorouracil (FdUMP) prevents DNA synthesis due to a lack of dTTP, and causes what has been called a "thymineless death" of the cell. The similarity of structures of FdUMP and the normal substrate dUMP suggests competitive inhibition. The entry of FdUMP into the enzyme active site is due to the similarity of its structure to that of the substrate, but a ternary complex forms that involves the coenzyme tetrahydrofolate and results in the covalent binding of FdUMP to the enzyme. Because of this, FdUMP ends up being an irreversible inhibitor. (*Biochemistry*, 2nd Edition, Moran & Scrimgeour, p. 22.19, 1994.)

3. (a) <u>Coenzyme components</u>. The coenzymes $NAD^{\oplus}$, FAD, and coenzyme A contain nucleotide components.

(b) <u>Nucleic acid components</u>. Nucleoside triphosphates are required for DNA and RNA synthesis.

(c) <u>Energy sources</u>. ATP represents the major form of stored energy in the cell. It is needed for muscular movement and certain metabolic processes.

(d) <u>Allosteric control</u>. ATP, ADP, and other nucleotides serve as allosteric modulators in several metabolic pathways.

(e) <u>Group donors</u>. UDP-glucose, CDP-choline, and S-adenosylmethionine are nucleotide derivatives (metabolite coenzymes) that serve as carriers of groups to be donated to other molecules in various reactions.

(f) <u>Signal transmission</u>. cAMP and cGMP serve as second messengers in transmission of extracellular signals intracellularly.

4. AZT is a thymidine analog. Nucleosides must be phosphorylated to be active in cells (eg. for nucleic acid synthesis). AZT is phosphorylated to AZT triphosphate by the action of kinases normally present in the cell. AZT triphosphate inhibits HIV replication by inhibition of the HIV-DNA polymerase and has a much less pronounced effect on the host cell's own DNA polymerase.

5. These observations are consistent with a ping-pong mechanism in which covalent catalysis occurs. ATP binds to the enzyme active site and transfers a phosphoryl group to the enzyme (covalent catalysis) and the product ADP is then released. When the second substrate (CDP) is added, it receives the phosphoryl group held by the enzyme, thus forming the second product, CTP.

CHAPTER 10

A. True–False

1. <u>True.</u> Phosphatidyl choline (lecithin) is a glycerophospholipid, an amphipathic molecule.

2. <u>False.</u> Although glycerophospholipids and most sphingolipids are ionic, cholesterol (a sterol) is uncharged.

3. <u>False.</u> They are similar to phosphatidates but contain a hydrocarbon chain attached to carbon-1 of the glyceryl backbone by a vinyl ether linkage.

4. <u>False.</u> Some specifically hydrolyze only fatty acyl ester bonds of phospholipids. See text Figure 10.9 for hydrolytic specificities of the phospholipases.

5. <u>False.</u> It is connected to sphingosine via an amide bond.

6. <u>True.</u> They contain one or more of the NeuNAc moieties.

7. <u>False.</u> In aqueous solution, amphipathic molecules aggregate to form micelles (see Chapter 2). Cholesterol contains a hydroxyl group but does not have sufficient amphipathic character to form micelles.

8. <u>True.</u> Even though they are polar, water molecules are not ionized and diffuse across membranes extremely rapidly.

9. <u>True.</u> They are triesters and have no ionizable groups.

10. <u>False.</u> Peripheral proteins are on one membrane surface or the other, but integral proteins are embedded in the lipid bilayer and extend through the interior of the membrane.

11. <u>True.</u> The membrane is asymmetric with respect to the proteins, the lipids, and the carbohydrate moieties (if any) that it contains.

12. <u>True.</u> Evidence of such movement led, in part, to the formulation of the fluid mosaic model.

B. Short Answers

1. glycerophospholipids; sphingolipids; cholesterol

2. phosphate

3. serine; choline; ethanolamine

4. Gangliosides

5. Cerebrosides (or galactocerebrosides)

6. phospholipase D

7. Cerebrosides

8. Flippase

9. liquid-crystalline phase; gel phase

10. flip-flop (or transverse diffusion)

11. peripheral; integral

C. Problems

1. (a) Glucocerebroside yields sphingosine, a fatty acid, and glucose.

   (b) Sphingomyelin yields sphingosine, a fatty acid, a phosphate, and choline.

   (c) Lecithin yields glycerol, phosphate, choline, and two fatty acids.

2. Cholesterol molecules associate with the nonpolar tails of the phospholipids such that the hydroxyl group of cholesterol is at the aqueous surface of the lipid layer. The nearly planar cholesterol molecule lies approximately parallel to the fatty acid chains of surrounding lipids.

3. Eicosanoids are derivatives of polyunsaturated, 20-carbon fatty acids such as arachidonic acid (20:4). Prostaglandins are one type of eicosanoid. Prostaglandin E_2 constricts blood vessels. Thromboxane A_2 is involved in blood clotting. Leukotriene D_4 mediates smooth muscle contraction. (See text Figure 10.19 for structures.)

4. The change in entropy of the water molecules provides the driving force. As water molecules orient themselves around a mono- or bilayer of lipids, the system becomes more ordered. Stabilization also occurs due to hydrophobic interactions of the nonpolar lipid tails.

5. Carbohydrates are always found on the exterior surface of the cell membrane. They may be attached either to lipid molecules or to proteins.

6. Apparently $Ca^{2\oplus}$ and $Mg^{2\oplus}$ ions help to hold the peripheral proteins to the membrane surface via charge-charge interactions. Chelating agents complex these ions and disrupt the interactions between the peripheral proteins and the membrane. The peripheral proteins are thus freed from the membrane.

7. Primary active transport requires some direct source of energy such as light, ATP, or electron transport. The text example is that of bacteriorhodopsin, which uses light energy to generate a proton concentration gradient. Secondary active transport is driven by an ion concentration gradient. For example, the flow of protons into an *E. coli* cell down a concentration gradient allows lactose transport into the cell against its concentration gradient.

8. Most integral proteins appear to have one or more segments of nonpolar amino acid residues that span the nonpolar center of the lipid bilayer. Other areas of the protein are located on or near the membrane surfaces.

9. The triacylglycerol is quite hydrophobic with no charges or polar groups and is, therefore, most soluble in the chloroform component of the solvent. (The nonpolar solvents are the "moving phase" on silica gel.) Lecithin is a phosphatidate which has a charged polar "head" region and is less soluble in the moving nonpolar phase and does not, therefore, travel very far from the origin.

10. Unsaturated fatty acid chains have *cis* double bonds that produce a bend in the chain and prevent close packing of the fatty acyl chains, thus providing a more fluid environment. Also, double bonds are a bit shorter than single bonds and that causes a little less interaction between the unsaturated hydrocarbon chains due to the shorter length. Cholesterol broadens the phase-transition temperature range of the bilayer. It intercalates between lipid molecules, which causes decreased fluidity due to restricted mobility of adjacent fatty acyl chains. It increases fluidity by disrupting the ordered packing of fatty acyl chains of the lipid molecules. The net effect of cholesterol in membranes is that it helps maintain fairly constant fluidity despite fluctuations in temperature or degree of fatty acid saturation.

11. Endocytosis: A macromolecule binds to a protein receptor on the extracellular surface of the cell membrane. The membrane then invaginates, forming a lipid vesicle from part of the membrane. Exocytosis: Materials to be excreted from the cell are packaged in lipid vesicles that bud off from the Golgi apparatus. These vesicles fuse with the cell membrane and release their contents into the extracellular space.

D. Additional Problems

1. The number of molecules in one surface of the bilayer is:
$$2.86 \times 10^4 \text{ molecules/2 surfaces} = 1.43 \times 10^4 \text{ molecules/surface.}$$
The surface area of one face is 100. μm^2 or 1.00×10^8 nm^2. The surface area per molecule is:
$$1.00 \times 10^8 \text{ nm}^2/\text{face} \div 1.43 \times 10^4 \text{ molecules/face} = 6.99 \times 10^3 \text{ nm}^2/\text{molecule.}$$

2. Using 4.5 nm as the lipid bilayer thickness, as shown in Figure 10.1, the total contribution of the bilayer to the thickness of the liposome is:
$$2 \times 4.5 \text{ nm} = 9.0 \text{ nm.}$$
The diameter of the inner aqueous sphere is then $(40.0 - 9.0)$ nm $= 31.0$ nm. The volume of a sphere is $4/3 \, \pi r^3$. Thus, the volume is:
$$(4/3)(3.14)(15.5 \text{ nm})^3 = 1.56 \times 10^4 \text{ nm}^3.$$

3. $s = (4Dt)^{1/2}$; $s^2 = 4Dt$, so $t = s^2/4D$.
If $s = 2.0$ μm or 2.0×10^{-4} cm, then
$$t = (4.0 \times 10^{-8} \text{ cm}^2)/(4)(10^{-8} \text{ cm}^2/\text{sec}) = 1.0 \text{ sec.}$$

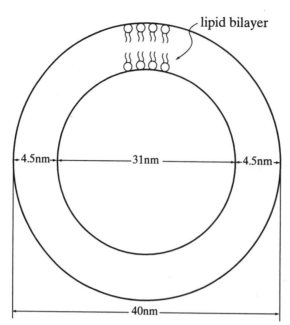

Figure 10.1 Cross-section of a Liposome.

4. Lipids with a free -NH$_2$ group, such as phosphatidyl serine and phosphatidylethanolamine, react with TNBS. Figure 10.2 illustrates the general reaction. Figure 10.3 illustrates the mechanism.

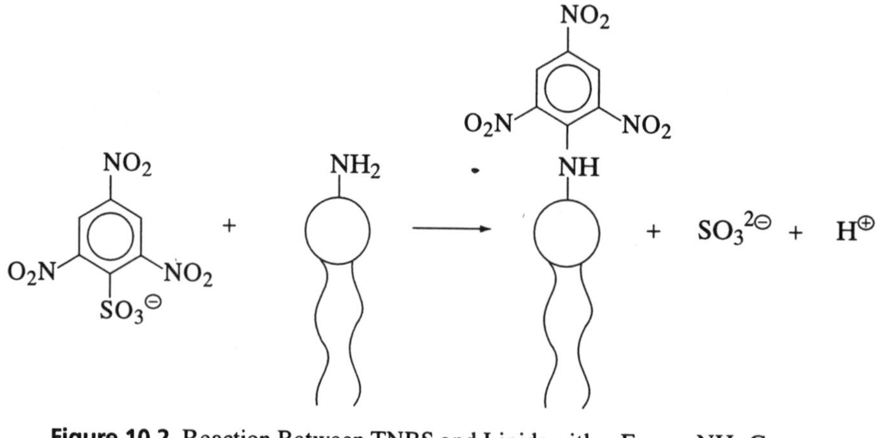

Figure 10.2 Reaction Between TNBS and Lipids with a Free —NH$_2$ Group.

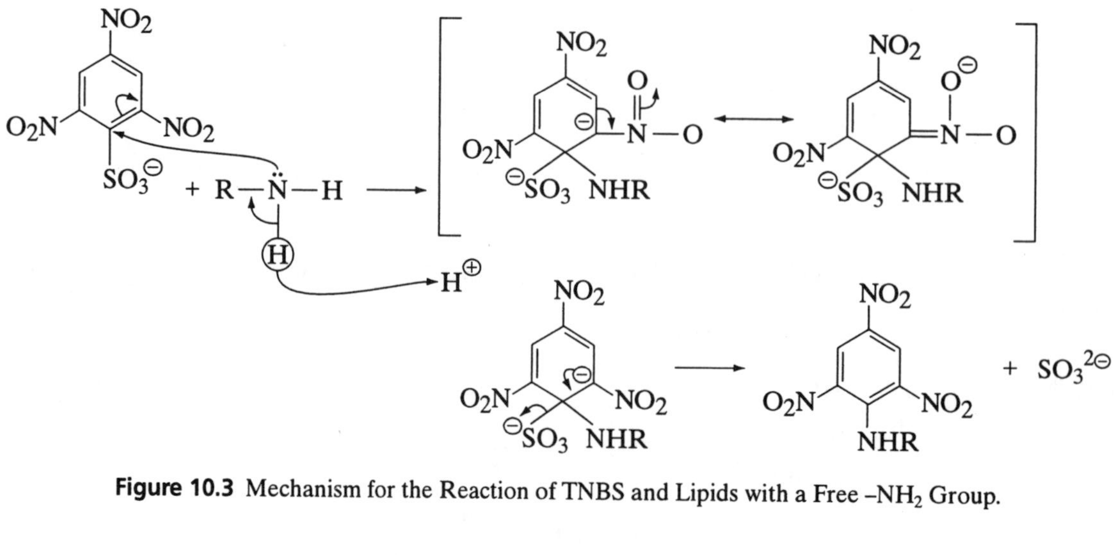

Figure 10.3 Mechanism for the Reaction of TNBS and Lipids with a Free –NH$_2$ Group.

5. An α-helix incorporates 3.6 amino acid residues per turn and covers a distance of 0.54 nm. The average number of residues required to traverse the membrane in one direction is:

$$[4 \text{ nm} \times 1 \text{ turn}/0.54 \text{ nm}][3.6 \text{ residues/turn}]$$

$$= 26.7 \text{ residues (per traversal)}.$$

Since 73% of the amino acid residues are in α-helix areas within the membrane, there are 247 residues × 0.73 = 180 residues in these areas. Therefore, the number of traversals is:

$$180 \text{ residues} \times 1 \text{ traversal}/26.7 \text{ residues} = 7 \text{ traversals}.$$

CHAPTER 11

A. True–False

1. <u>True.</u> The two major divisions of metabolism are degradative (catabolic) and synthetic (anabolic).

2. <u>False.</u> The flux of material through a pathway is controlled by availability of reactants, allosteric control, and covalent modification of interconvertible enzymes. Pathway flux can be increased or decreased as required by the cell.

3. <u>False.</u> Energy transfer efficiency does not approach 100%. In general, it is less than 50%.

4. <u>True.</u> Entropy is a measure of randomness or disorder. The NaCl is in a much less ordered state (greater randomness) in solution than in the very ordered crystalline state.

5. <u>True.</u> When ΔG is negative, the reaction can proceed without an external supply of energy. Note that the sign of ΔG, not ΔG$^{0'}$, is the indicator of spontaneity.

6. <u>False.</u> Other factors are also involved in determining whether a process is spontaneous. See the next question.

7. <u>False.</u> The change in entropy, ΔS, is also a factor. The relationship is: ΔG = ΔH – TΔS.

8. <u>False.</u> Cellular reactions occur at a steady state in which substrate is being supplied at about the same rate that product is being removed.

9. <u>True.</u> Such reactions can serve as control points for pathways.

10. <u>False.</u> It indicates that, at equilibrium, there will be less of C and D than of A and B. The reverse reaction is actually more favored.

11. <u>True.</u> The phosphoryl-group-transfer potential of ATP is such that ADP can accept a phosphoryl group from some compounds to form ATP, which can donate a phosphoryl group to other compounds.

B. Short Answers

1. anabolic; catabolic

2. linear; cyclic; spiral

3. increase; release

4. glycolysis

5. phosphoanhydride bond

6. reducing agent; oxidized

7. phosphagens; phosphocreatine; phosphoarginine

8. energy-rich compounds

9. product; feedback inhibition; substrate; feed-forward activation

10. reduction potential

C. Problems

1. Yes, it is possible under appropriate non-standard conditions. While the change in standard free energy, ΔG$^{0'}$, may be positive, it is the free energy change, ΔG, for the non-standard set of conditions that must be negative for the reaction to proceed spontaneously.

2. In the expression that relates ΔG and $\Delta G^{0\prime}$,

$$\Delta G = \Delta G^{0\prime} + RT \ln [C][D]/[A][B],$$

the ln term is the only factor that can change. If the mass action ratio, Q, is less than one, which is the case when concentrations of reactants are greater than concentrations of products, the ln term will be negative. Still, the negative term must be larger than the positive $\Delta G^{0\prime}$ value for ΔG to also be negative.

3. Use the relationship:

$\Delta G^{0\prime} = -RT \ln K_{eq}$, and solve for $\ln K_{eq}$.

$\ln K_{eq} = \Delta G^{0\prime}/(-RT)$.

From text Table 11.3, $\Delta G^{0\prime}$ for this reaction is –21 kJ/mol. Using the value from the text for R and the standard temperature of 25°C converted to Kelvins,

$$\ln K_{eq} = \frac{(-21 \text{ kJ/mol})(10^3 \text{ J/kJ})}{(-8.315 \text{ J K}^{-1} \text{ mol}^{-1})(298 \text{ K})} = 21 \times 10^3 / 2.48 \times 10^3 = 8.47$$

$K_{eq} = e^{8.47} = 4758 = 4.8 \times 10^3$.

4. Characteristically, phosphorylation activates the interconvertible enzymes of catabolic pathways, but those in anabolic pathways are usually activated by dephosphorylation.

5. Compartmentation allows (1) separate pools of metabolites, (2) opposing metabolic pathways to operate simultaneously, (3) high local concentrations of metabolites, (4) coordinated regulation of multiple enzymes, and (5) site-specific regulation of metabolic processes by specialized tissues.

6. The formation of one ATP from ADP and P_i requires 30. kJ/mol, the same quantity liberated when ATP is hydrolyzed, as indicated in text Table 11.1. Therefore, 32 moles of ATP × 30. kJ/mol ATP = 960 kJ. The in vitro production of 686 kcal/mol × 4.184 kJ/kcal = 2870 kJ. The percentage of energy captured from the glucose is:

$$\% \text{ capture} = \frac{960 \text{ kJ} \times 100}{2870 \text{ kJ}} = 33.45 = 33\%.$$

7. $\Delta G = \Delta G^{0\prime} + RT \ln Q$, where Q is the mass action ratio of products to reactants, $\dfrac{[\text{glucose}][P_i]}{[\text{glucose 6 – phosphate}]}$.[1]

$$\Delta G = -14 \text{ kJ/mol} + (8.31 \times 10^{-3} \text{ kJ/mol K})(310. \text{ K}) \ln \frac{[2 \times 10^{-4}][5 \times 10^{-2}]}{[10^{-3}]}$$

$$= -14 \text{ kJ/mol} + (-11.86 \text{ kJ/mol}) = -25.86 = -26 \text{ kJ/mol}.$$

8. (a) Assume the cell might use the energy supplied by the hydrolysis of ATP to AMP and PP_i to drive the reaction. This might be accomplished by using ATP to donate the AMP group to the carboxylic acid and then using the activated acid to react with the thiol.

$$\text{R-COO}^\ominus + \text{ATP} \longrightarrow \text{R-CO-AMP} + PP_i$$
$$\text{R-CO-AMP} + \text{R}\prime\text{-SH} \rightarrow \text{R-CO-SR}\prime + \text{AMP}$$

The sum of these reactions is:

$$\text{R-COO}^\ominus + \text{ATP} + \text{R}\prime\text{-SH} \rightarrow \text{R-CO-SR}\prime + \text{AMP} + PP_i$$

Notice that this is the same net reaction obtained by adding the following processes for which the standard free energies are known:

$$\text{R-COO}^\ominus + \text{R}\prime\text{-SH} \rightarrow \text{R-CO-SR}\prime + H_2O \quad \Delta G^{0\prime} = 10. \text{ kJ/mol}$$
$$\text{ATP} + H_2O \longrightarrow \text{AMP} + PP_i \quad \Delta G^{0\prime} = -32 \text{ kJ/mol}$$

(b) The net $\Delta G^{0\prime}$ for this combination is –32 kJ/mol + 10. kJ/mol = –22 kJ/mol. Since this standard free energy change is negative, the reaction should proceed spontaneously. An additional driving force might be the hydrolysis of PP_i to 2 P_i, catalyzed by pyrophosphatase. The $\Delta G^{0\prime}$ for this reaction is –34 kJ/mol, which, added to the previous total,

---

[1] The concentration of water as a reactant is not included in the ratio, Q, because it is essentially constant. The small amount used up in the reaction is negligible compared to that available in the surrounding medium, the cell cytosol. In addition, the cellular water concentration is readily maintained, since plasma membranes are freely permeable to water.

would give an overall total of –56 kJ/mol. (Note that the cell would have to have the appropriate enzymes to catalyze these steps.)

9. Calculate the standard free energy change from the relationship: $\Delta G^{0'} = -nF\Delta E^{0'}$. Use the standard reduction potentials from text Table 11.4 to calculate the change in standard reduction potentials, $\Delta E^{0'}$, using this relationship:

$$\Delta E^{0'} = E^{0'}_{red} - E^{0'}_{ox}$$

(Note that $_{red}$ and $_{ox}$ are used to indicated the half-reaction involving reduction and oxidation, respectively.) From text Table 11.4,

$$Q + 2\,H^{\oplus} + 2\,e^{\ominus} \rightarrow QH_2 \qquad E^{0'} = 0.04\ V$$
$$Fumarate + 2\,H^{\oplus} + 2\,e^{\ominus} \rightarrow succinate\ \ E^{0'} = 0.03\ V$$

For the reaction,

$$succinate + Q \rightarrow fumarate + QH_2,$$
$$\Delta E^{0'} = 0.04\ V - 0.03\ V = 0.01\ V.$$

Use this change in standard reduction potentials in the equation given above.

$$\Delta G^{0'} = -(2)(96.48\ kJ/V\ mol)(0.01\ V) = -1.9296 = -2\ kJ/mol.$$

10. Combination of the two steps yields a negative $\Delta G^{0'}$ (–3.0 kJ/mol) for the overall process, indicating that the process is energy-releasing and, therefore, spontaneous.

D. Additional Problems

1. (a) Use the Henderson-Hasselbalch equation:

$pH = pK_a + \log\,[A^{\ominus}]/[HA]$

where $A^{\ominus} = ATP^{4\ominus}$, and $HA = HATP^{3\ominus}$.

$\log\,[A^{\ominus}]/[HA] = pH - pK_a,$

$\log\,[ATP^{4\ominus}]/[HATP^{3\ominus}] = 7.40 - 6.95 = 0.45.$

$[ATP^{4\ominus}]/[HATP^{3\ominus}] = 10^{0.45} = 2.8$ or a ratio of 2.8 to 1.

If the ratio is 2.8/1, the percentage of $ATP^{4\ominus}$ is:

$\%\ ATP^{4\ominus} = [2.8/(2.8 + 1)] \times 100 = 73.68 = 74\%.$

(b) Doing the same calculation using pH 7.0 rather than 7.40 gives a percentage of 52% for the $ATP^{4\ominus}$ form.

2. (a) For the reaction:

$$Glucose + ATP \rightarrow Glucose\ 6\text{-phosphate} + ADP,$$

the change in standard free energy is related to the equilibrium constant by:

$\Delta G^{0'} = -RT \ln K_{eq}.$

$\ln K_{eq} = \Delta G^{0'}/-RT$

$$\ln K_{eq} = \frac{-16.7\ kJ/mol}{(-8.31 \times 10^{-3}\ kJ/mol\ K)(310.\ K)}$$

$$\ln K_{eq} = \frac{16.7\ kJ/mol}{2.576\ kJ/mol} = 6.48$$

$$K_{eq} = e^{6.48} = 645.$$

This value and the negative standard free energy change indicate that the reaction should proceed spontaneously to the right.

(b) Use a series of approximations to determine the equilibrium concentrations. At equilibrium, [glucose 6–phosphate] = [ADP] = x, and [Glucose] = [ATP] = (0.0500 –x) M. The large K_{eq} value indicates that, at equilibrium, the concentrations of products are much greater than the concentrations of remaining reactants. Trial 1: If 90.% of reactants were consumed, x = (0.0500 M)(0.90) = 0.0450 M. The value of K_{eq} would be $(0.0450)^2/(0.00500)^2 = 81$, which is smaller than the known value of K_{eq}, indicating that greater than 90.% of the reactants must have reacted. Trial 2: If 98% of reactants were consumed, x = (0.0500 M)(0.98) = 0.0490 M. The value of K_{eq}, determined as in Trial 1 is

2,400. Since this is much larger than the known value of K_{eq}, less than 98% of the reactants reacted. Similar trials show that about 96.2% of the reactants were consumed. Thus, the concentrations of glucose 6-phosphate and ADP are about 0.0481 M and those of glucose and ATP are about 0.00190 M.

3. $\Delta G = \Delta G^{0\prime} + RT \ln \dfrac{[\text{glucose 6-phosphate}][\text{ADP}]}{[\text{glucose}][\text{ATP}]};$

$$\ln \frac{[\text{glucose 6-phosphate}][\text{ADP}]}{[\text{glucose}][\text{ATP}]} = \frac{\Delta G - \Delta G^{0\prime}}{RT};$$

$$\ln \frac{[\text{glucose 6-phosphate}][\text{ADP}]}{[\text{glucose}]} + \ln \frac{[\text{ADP}]}{[\text{ATP}]} = \frac{\Delta G - \Delta G^{0\prime}}{RT};$$

$$\ln \frac{[\text{glucose 6-phosphate}]}{[\text{glucose}]} = \frac{\Delta G - \Delta G^{0\prime}}{RT} + \ln \frac{[\text{ATP}]}{[\text{ADP}]};$$

$$\ln \frac{[\text{glucose 6-phosphate}]}{[\text{glucose}]} = \frac{-33.9 \text{ kJ/mol} - (-16.7 \text{ kJ/mol})}{(8.31 \times 10^{-3} \text{ kJ/mol K})(310 \text{ K})} + \ln \frac{[13]}{[1]};$$

$$\ln \frac{[\text{glucose 6-phosphate}]}{[\text{glucose}]} = \frac{-17.2}{2.576} + 2.56 = -6.69 + 2.56 = -4.13;$$

$$\frac{[\text{glucose 6-phosphate}]}{[\text{glucose}]} = e^{-4.13} = 0.0161 = 0.016.$$

The ratio of glucose to glucose 6-phosphate is the reciprocal of 0.016 or about 63 to 1.

4. $\Delta G = \Delta G^{0\prime} + RT \ln K_{eq};$

$$\ln K_{eq} = \frac{\Delta G - \Delta G^{0\prime}}{RT}$$

$$\ln \frac{[\text{ADP}][\text{P}_i]}{[\text{ATP}]} = \frac{\Delta G - \Delta G^{0\prime}}{RT}$$

$$\ln \frac{[\text{ATP}][\text{P}_i]}{[\text{ADP}]} = \frac{-41.8 \text{ kJ/mol} - (-30 \text{ kJ/mol})}{(8.31 \times 10^{-3} \text{ kJ/mol K})(310 \text{ K})} = -4.58$$

$$\frac{[\text{ATP}][\text{P}_i]}{[\text{ADP}]} = e^{-4.58} = 0.0102.$$

Let $x = [\text{ADP}] = [\text{P}_i]$. When $[\text{ATP}] = 1$, $x = (1 \times 0.0102)^{1/2} = 0.101$. The ratio of ATP to ADP is then 1/0.101 or 9.9 to 1. Since the ΔG for this reaction is even more negative than the standard free energy change, $\Delta G^{0\prime}$, the tendency for hydrolysis is even greater than at standard conditions.

5. (a) Both systems produce the same amount of stored energy as ATP (2 moles each), but glycolysis produces 2 NADH which represents reducing equivalents capable of producing more ATP by their oxidation in the electron transport system.

(b) No net oxidation of glucose occurs when lactate is formed, since there are no reduced cofactors formed. Oxidation does occur when pyruvate is formed as noted by the 2 NADH formed.

(c) The system in which lactate is formed has the larger negative value of standard free energy change and thus has the greater tendency to proceed spontaneously as written.

CHAPTER 12

A. True–False

1. <u>False.</u> They are in the cytosol of eukaryotic cells.

2. <u>True.</u> The change in $\Delta G^{0\prime}$ for the overall process is negative. Some of the energy derived from glycolytic catabolism of glucose is stored in ATP, formed in two pathway reactions.

3. <u>True.</u> Considering pyruvate the end product of glycolysis, oxidation is achieved without the direct use of oxygen.

4. <u>True.</u> In yeast, pyruvate is decarboxylated to acetaldehyde, which is then reduced to ethanol.

5. <u>True.</u> The enzyme pyruvate decarboxylase is absent in humans.

6. <u>False.</u> This kind of intramolecular transfer is characteristic of enzymes in category 2, transferases.

7. <u>True.</u> Combination of the two stages results in a net production of two ATP per mole of glucose.

8. <u>True.</u> Entry of substrates as either glucose or as fructose 6-phosphate can be controlled by regulating these two reactions.

9. <u>False.</u> Glucose is phosphorylated after entering the cell, which prevents it from passing back through the cell membrane.

10. <u>True.</u> One mole of either fructose, galactose, mannose, or glucose produces a net of two moles of ATP via glycolysis.

B. Short Answers

1. isozymes; phosphoenolpyruvate

2. phosphatases

3. fructose 1,6-*bis*phosphate; ATP

4. hexokinase; phosphofructokinase-1; pyruvate kinase

5. substrate-level phosphorylation

6. epimerase

7. the Pasteur effect

8. a deficiency of the enzyme lactase

9. citrate; ATP

10. pyruvate and ATP

C. Problems

1. D-glucose + 2 ADP + 2 P_i + 2 $NAD^{\oplus}$ → 2 pyruvate + 2 ATP + 2 NADH + 2 $H^{\oplus}$ + 2 H_2O

2. Glycolysis would not proceed under these conditions because ATP is required in the first half of glycolysis at two separate steps. (Which steps are they?)

3. Lactate is not constantly formed in mammalian cells. In most tissues, pyruvate is normally decarboxylated to yield acetyl CoA that fuels the Krebs cycle, under aerobic conditions. In tissues such as those of exercising muscle, in which oxygen is in low supply, some pyruvate is converted to lactate to keep glycolysis going by regenerating $NAD^{\oplus}$ (required for which step?).

4.

Hexose	Initial Glycolytic Intermediate	Number of Moles of Intermediate
mannose	fructose 6-phosphate	1
fructose	glyceraldehyde 3-phosphate and dihydroxyacetone phosphate	1 1
galactose	glucose 6-phosphate	1

5. If not for triose phosphate isomerase, only one half of the glucose molecule, D-glyceraldehyde 3-phosphate, would be utilized.

6. Figure 12.2 shows the regeneration of NAD$^\oplus$ in muscle cells contrasted to that in yeast cells.

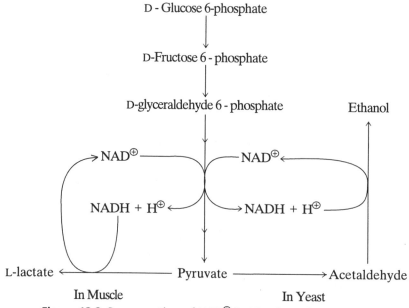

Figure 12.2 Regeneration of NAD$^\oplus$ in Muscle and Yeast Cells.

7. Enolase catalyzes the conversion of 2-phosphoglycerate to phosphoenolpyruvate. In the presence of an inhibitor, the substrate 2-phosphoglycerate would accumulate, as would other intermediates formed in previous, near-equilibrium reactions in the pathway.

8. Dihydroxyacetone phosphate would accumulate and present a problem unless disposed of by some other mechanism. Also, the individual would obtain only half the energy from glucose that a normal individual would obtain, since only the glyceraldehyde 3-phosphate half of the cleaved fructose 1,6-*bis*phosphate molecule would be available for energy production.

9. Figure 12.3 illustrates the reaction between iodoacetate and the cysteine sulfhydryl group of the enzyme. This is irreversible inhibition.

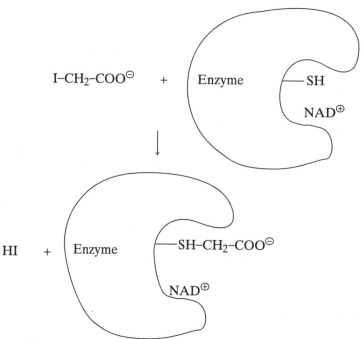

Figure 12.3 Iodoacetate Inhibition of Glyceraldehyde 3–Phosphate Dehydrogenase.

10. ATP supplies phosphoryl groups in two reactions in the hexose stage. Inorganic phosphate, P_i, is the source of the other phosphoryl groups. One mole of P_i is used per mole of glyceraldehyde 3-phosphate.

D. Additional Problems

1. Use Table 12.1 and combine the standard free energy changes for reactions 1, 2, 3, 4, and 5. To that value, combine two times the standard free energy changes for reactions 6 through 10, since these reactions occur twice in consuming the two halves of the original glucose molecule.

$$\Delta G^{0\prime} = [-16.7 + 1.67 + (-14.2) + 24.0 + 7.66] + 2[6.28 + (-18.8) + 4.44 + 1.84 + (-31.4)] \text{ kJ/mol}$$
$$= 2.4 + 2(-37.6) = -72.8 \text{ kJ/mol}$$

Table 12·1 Standard Free Energy Changes for Glycolysis

Step No.	Reaction	Enzyme	$\Delta G^{0\prime}$, kJ/mol
1	Glucose → G6P	hexokinase	−16.7
2	G6P → F6P	glucose 6-phosphate isomerase	+1.67
3	F6P → F1,6bisP	phosphofructokinase–1	−14.2
4	F1,6bisP → DHAP + G3P	aldolase	+24.0
5	DHAP → G3P	triose phosphate isomerase	+7.66
6	G3P → 1,3bisPG	glyceraldehyde 3–phosphate dehydrogenase	+6.28
7	1,3bisPG → 3PG	phosphoglycerate kinase	−18.8
8	3PG → 2PG	phosphoglycerate mutase	+4.44
9	2PG → PEP	enolase	+1.84
10	PEP → pyruvate	pyruvate kinase	−31.4

2. Glycolysis produces a net of two ATP per glucose utilized. Each ATP represents −30. kJ/mol, so the total energy from glycolysis is equal to that tabulated in the previous problem, −72.8 kJ/mol, plus that captured as ATP, $2 \times (-30.\text{ kJ/mol})$.

Total energy = −72.8 kJ/mol + (−60. kJ/mol) = −132.8 = −133 kJ/mol.

The %efficiency of energy capture is:

$$\% \text{ efficiency} = \frac{60.\text{ kJ/mol} \times 100}{-133 \text{ kJ/mol}} = 45\%$$

3. Formation of two lactates from the two pyruvates formed from one glucose represents $2 \times (-25.1 \text{ kJ/mol})$ or −50.2 kJ/mol. The total energy released is −133 kJ/mol + (−50.2 kJ/mol) = −183.2 or −183 kJ/mol. The percent efficiency is:

$$\% \text{ efficiency} = \frac{2\,(-30.\text{ kJ/mol} \times 100}{-183 \text{ kJ/mol}} = 33\%$$

This indicates that the efficiency is less than that of glycolytic catabolism of glucose that does not include the conversion of pyruvate to lactate.

4. The large negative standard free energy change for pyruvate formation (step 10), −31.4 kJ/mol, helps drive glycolysis as a whole. The glycolytic process has a $\Delta G^{0\prime}$ of −72.8 kJ/mol as determined in problem D.1. Under physiological conditions, the change in free energy, ΔG, is even more negative than is $\Delta G^{0\prime}$.

As product is formed in the aldolase reaction (step 4), it is used in the next pathway step. This prevents a disadvantageous equilibrium from being established at the aldolase step.

5. Enolase catalyzes the conversion of 2-phosphoglycerate to phosphoenolpyruvate (step 9). With enolase deficiency, 2-phosphoglycerate accumulates, as do glycolytic intermediates of previous steps, including 1,3-*bis*phosphoglycerate. 2,3-*Bis*phosphoglycerate modulates the affinity of hemoglobin for oxygen. The greater the concentration of 2,3-*bis*phosphoglycerate, the lower the affinity of hemoglobin for oxygen, which results in less oxygen delivered to the tissues. Text Figure 12.30 illustrates the formation and degradation of 2,3-*bis*phosphoglycerate, which involves the nonglycolytic enzymes *bis*phosphoglycerate mutase and 2,3-*bis*phosphoglycerate phosphatase, respectively. A buildup of 1,3-*bis*phosphoglycerate results in the formation of more 2,3-*bis*phosphoglycerate.

CHAPTER 13

A. True–False

1. <u>True.</u> FAD is a prosthetic group of an enzyme of the succinate dehydrogenase complex that accepts hydrogens from succinate and immediately transfers them to coenzyme Q, the mobile carrier of reducing power within the inner mitochondrial membrane.

2. <u>True.</u> The citric acid cycle generates six molecules of CO_2 for each molecule of glucose catabolized.

3. <u>False.</u> Your text points out that the cycle is a multistep catalyst. Since the process is cyclic, the intermediates are continually recycled. However, sufficient amounts of oxaloacetate must be present to allow appropriate rates of acetyl CoA uptake.

4. <u>False.</u> There are two six-carbon intermediates, one five-carbon intermediate, and five four-carbon intermediates as indicated in Figure 13.1. To better familiarize yourself with these intermediates, make a list of them by name.

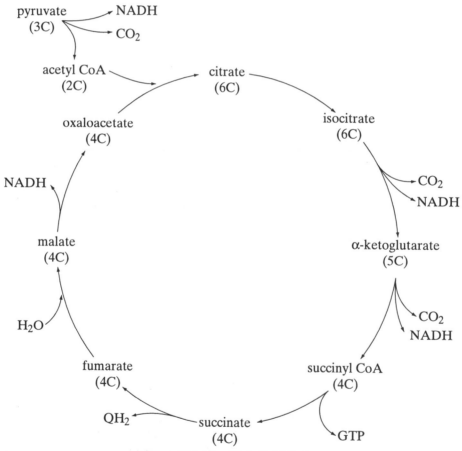

Figure 13.1 The Citric Acid Cycle.

5. <u>False.</u> They are located in mitochondria.

6. <u>True.</u>

7. <u>True.</u> Succinyl CoA condenses with glycine to initiate porphyrin biosynthesis, but succinyl CoA can be replenished by degradation of some amino acids.

8. <u>False.</u> Although coenzyme A is required in two different steps as pyruvate is catabolized, two moles of coenzyme A are formed as products in the citric acid cycle, so there is no net consumption of coenzyme A.

9. <u>False.</u> In addition to the availability of acetyl CoA and oxaloacetate to initiate citrate formation, control is exercised through allosteric modulators and their effects on the cycle enzymes citrate synthase, isocitrate dehydrogenase, and the α-ketoglutarate dehydrogenase complex.

10. <u>True.</u> Glycolysis produces two ATP per mole of glucose. From the two moles of pyruvate produced by glycolysis, the citric acid cycle produces two moles of GTP (or ATP).

11. <u>True.</u> The prochiral molecule contains a carbon of the type C_{aacd}, which can be converted to a chiral molecule by replacement of one of the *a* groups by a group different from the remaining three groups. The carbon would then be C_{abcd}.

B. Short Answers

1. coenzyme A, $NAD^{\oplus}$, lipoate, FAD, and TPP

2. succinate, fumarate, malate, and oxaloacetate (Prior to the elucidation of the citric acid cycle, biochemist Albert Szent-Gyorgyi discovered that these dicarboxylic acids stimulate respiration.)

3. CO_2; $NAD^{\oplus}$

4. the pyruvate dehydrogenase complex [This complex contains multiple copies of pyruvate dehydrogenase (E_1), dihydrolipoamide acetyltransferase (E_2), and dihydrolipoamide dehydrogenase (E_3).]

5. the α-ketoglutarate dehydrogenase complex

6. succinyl CoA; succinate

7. five (including the conversion of pyruvate to acetyl CoA); NADH; QH_2

8. malonate

9. α-ketoglutarate; succinyl CoA (as well as citrate and oxaloacetate)

10. NADH

11. L-malate; fumarate

C. Problems

1. Coenzyme A is composed of an ADP moiety, an esterified phosphate, a pantothenic acid unit, and a mercaptoethylamine unit. Coenzyme A functions as a carrier of acyl groups such as the acetyl group formed when pyruvate is oxidatively decarboxylated by the pyruvate dehydrogenase complex, and the succinyl group formed when α-ketoglutarate is oxidatively decarboxylated by the α-ketoglutarate dehydrogenase complex.

2. Sources of acetyl CoA are: pyruvate (from glucose and other sources); fatty acids (from triacylglycerols and other lipids); the carbon skeletons of some amino acids.

3. Pyruvate, produced in the cytosol by glycolysis, enters mitochondria through porin in the outer membrane and by the aid of pyruvate translocase in the inner membrane. In the mitochondrial matrix, pyruvate is converted to acetyl CoA, a substrate for the citric acid cycle.

4. A kinase and a phosphatase are part of the mammalian pyruvate dehydrogenase complex. Pyruvate dehydrogenase kinase catalyzes the phosphorylation of E_1, pyruvate dehydrogenase, causing its inactivation. Pyruvate dehydrogenase phosphatase catalyzes the dephosphorylation and activation of E_1. Control of the complex is achieved by control of E_1, since it catalyzes the rate-determining step in the conversion of pyruvate to acetyl CoA and CO_2.

5. (a) An abundance of ADP stimulates isocitrate dehydrogenase so that more ATP is ultimately formed from the reduced cofactors NADH and QH_2 produced by the cycle. ADP also inhibits pyruvate dehydrogenase kinase, allowing pyruvate dehydrogenase to remain active.

   (b) Available acetyl CoA cannot all be converted to citrate when oxaloacetate levels are too low. More oxaloacetate is formed by the carboxylation of available pyruvate.

6. (a) four NADH, one QH_2, one GTP

   (b) three NADH, one QH_2, one GTP

   (c) eight NADH, two QH_2, two GTP (twice that of pyruvate since two moles of pyruvate are produced from each mole of glucose)

7.

$$
\begin{array}{l}
\text{CH}_2\text{—COO}^\ominus \\
\;\;| \\
\text{CH—COO}^\ominus \; + \; 2\,\text{NAD}^\oplus + \text{GDP} + \text{P}_i \longrightarrow 2\,\text{CO}_2 + 2\,\text{NADH} + \text{H}^\oplus + \text{GDP} + \text{CH}_2\text{—COO}^\ominus \\
\;\;| \qquad\qquad\qquad\qquad\qquad\qquad\qquad\qquad\qquad\qquad\qquad\qquad\qquad\qquad\qquad\quad | \\
\text{HO—CH—COO}^\ominus \qquad\qquad\qquad\qquad\qquad\qquad\qquad\qquad\qquad\qquad\qquad\qquad\qquad \text{CH}_2\text{—COO}^\ominus
\end{array}
$$

8. Pigeon muscle tissue contains citric acid cycle enzymes and cofactors but malonate inhibits the succinate dehydrogenase complex, causing succinate to accumulate.

9. Reactions in vivo are not at equilibrium, but are, instead, in a steady state of formation and consumption of intermediates. As isocitrate is formed in the cell, it is used in the next step of the citric acid cycle and more isocitrate is rapidly formed from citrate.

10. Accumulation of acetyl CoA partially reverses the reactions catalyzed by E_2 and E_3. Acetyl CoA also activates pyruvate dehydrogenase kinase in mammals. If sufficient product (acetyl CoA) of the step catalyzed by the complex is available, the enzyme is modulated to decrease pathway flux.

D. Additional Problems

1. Due to the pyruvate dehydrogenase deficiency, pyruvate and, consequently, lactate accumulate in the patient's serum, causing acidosis. Although less than normal amounts of acetyl CoA are produced from pyruvate, fatty acid oxidation yields acetyl CoA, as do some of the amino acids. Acetyl CoA from these sources fuels the citric acid cycle. Lowering carbohydrate levels in the diet would help minimize the pyruvate and lactate accumulation in the blood.

2. The labeled acetyl CoA from fatty acid catabolism condenses with oxaloacetate to form citrate. Two molecules of CO_2 are subsequently lost from citrate in the citric acid cycle but the carboxyl groups lost as CO_2 are those from the original oxaloacetate and not from the acetyl CoA. Therefore, the labeled carbons are incorporated in the four-carbon intermediates including succinate and, ultimately, oxaloacetate. Since oxaloacetate is a precursor for glucose via gluconeogenesis, some of the labeled carbons can be found in glucose. There is no net synthesis of glucose since two carbons, as acetyl CoA, go into the cycle and two carbons, as CO_2, are produced.

3.

| hydroxypyruvate | C | A | B | D | pyruvate |

4. Pyruvate carboxylase catalyzes the carboxylation of pyruvate to form oxaloacetate, an anaplerotic reaction that replenishes oxaloacetate for use in the citric acid cycle and gluconeogenesis. In this diseased state, oxaloacetate cannot be replenished at normal levels. This results in lower than normal ATP production, due to diminished activity of the citric acid cycle, and lower than normal glycogen stores available for energy production, due to diminished gluconeogenesis. Therefore, the afflicted individual has limited capacity for prolonged exercise.

5. (a) $\Delta G^{0\prime} = -RT\ln K_{eq}$

$$= -(8.315 \times 10^{-3}\,\text{kJ mol}^{-1}\,\text{K}^{-1})(298\,\text{K})(\ln 5 \times 10^5)$$

$$= -32.5\,\text{kJ mol}^{-1}.$$

(b) The second reaction is essentially a reversal of the first, so the $\Delta G^{0\prime}$ for the formation of oxaloacetate and acetyl CoA from citrate and CoASH would be $+32.5$ kJ mol^{-1}. However, the hydrolysis of ATP to ADP and P_i is also involved. Text table 11.1 gives $\Delta G^{0\prime} = -30$ kJ mol^{-1} for this reaction. Combination of these two values gives $\Delta G^{0\prime} = +2.5$ kJ mol^{-1} for the overall reaction. Since $\Delta G^{0\prime} = -RT\ln K_{eq}$,

$\ln K_{eq} = \Delta G^{0\prime}/-RT$

$$= +2.5\,\text{kJ mol–1}/(-8.315 \times 10^{-3}\,\text{kJ mol–1 K–1})(298\,\text{K})$$

$$= -1.0$$

$K_{eq} = 0.37$

(c) The large equilibrium constant for the citrate formation reaction and its negative $\Delta G^{0\prime}$ suggest that the reverse process would not proceed spontaneously without energy input. In the reaction catalyzed by citrate lyase, energy is provided by the net hydrolysis of one high-energy bond of ATP. The ATP is required for synthesis of the thiol ester intermediate, citryl CoA. When citrate accumulates and some is transported to the cytosol, sufficient ATP is available and not as much citrate must be catabolized by the citric acid cycle. Some ATP, therefore, can be expended to lyse citrate, forming acetyl CoA, which can be used for synthesis of fatty acids that can be stored for future energy needs.

CHAPTER 14

A. True–False

1. **True.** When this occurs, the overall stoichiometry involves the input of six molecules of glucose 6-phosphate and the production of 12 NADPH, 6 CO_2, and the regeneration of five molecules of glucose 6-phosphate.

2. **False.** High activity (especially in rapidly dividing cells) of the pentose phosphate pathway is required to supply ribose 5-phosphate for nucleotide biosynthesis.

3. **True.** It is formed by the catalytic action of glycogen phosphorylase on glycogen.

4. **False.** Glycogen synthase does not catalyze the formation of the α-(1→6) linkages required for formation of branches. This is accomplished by the enzyme amylo-(1,4→1,6)-transglycosylase (the branching enzyme). In addition, glycogenin catalyzes primer extension.

5. **False.** While some reactions are reversals of glycolytic reactions, four different enzymes are required to bypass the three metabolically irreversible reactions of glycolysis.

6. **False.** Four more moles of ATP are required for gluconeogenesis from two moles of pyruvate than are formed by glycolysis.

7. **True.** It involves glycolysis in peripheral tissues and gluconeogenesis in the liver.

8. **True.**

9. **False.** It is produced in the oxidative stage.

10. **True.** Hence, normal blood glucose levels are maintained at all times, even at the expense of peripheral (muscle) tissue. (Note: The brain does utilize ketone bodies during starvation.)

B. Short Answers

1. NADPH

2. transaldolase

3. gluconeogenesis
   pentose phosphate pathway
   gluconeogenesis
   glycogen synthesis

4. limit dextrin

5. pyruvate carboxylase; phosphoenolpyruvate carboxykinase; fructose 1,6-*bis*phosphatase; glucose 6-phosphatase

6. glycogenin

7. insulin; glucagon; epinephrine

8. hormonal induction

9. lactate; alanine (and most other amino acids); glycerol

10. substrate cycle

C. Problems

1. Because the liver and intestine are perfused in series, food fuels absorbed by the intestine flow immediately through the liver, which regulates the distribution of dietary fuels to other tissues.

2. The sigmoidal shape of the curve in the absence of AMP indicates positive cooperativity of binding of the substrate P_i. AMP enhances the reaction rate by lowering the apparent K_m for P_i. Thus, glycogen phosphorylase appears to be allosterically regulated, with AMP serving as a positive modulator. (Note: Glycogen phosphorylase exists in two active forms, glycogen phosphorylase *a*, a phosphorylated dimer, and glycogen phosphorylase *b*, a less active, non-phosphorylated dimer. These are the R (active) forms of the enzyme, which can also exist in T (inactive) forms. Glycogen phosphorylase *b* is formed when AMP binds to the non-phosphorylated T form of the enzyme.)

3. The reactions bypassed are the three metabolically irreversible reactions of glycolysis. Since they are not reversible, they must be bypassed by these four reactions specific to gluconeogenesis.

4. Glycogenolysis results in the production of glucose 1-phosphate and thence glucose 6-phosphate in both tissues. In muscle, the glucose 6-phosphate is metabolized for energy production via glycolysis and the citric acid cycle. In liver, most of the glucose 6-phosphate is converted to glucose, which leaves the cells and is delivered to other tissues, such as brain, for their energy needs.

5. Glycogen synthase catalyzes the addition of glucose residues from UDP-glucose to the nonreducing end of a glycogen chain. The branching enzyme, amylo-(1,4→1,6)-transglycosylase, catalyzes the removal of an oligosaccharide segment of at least six residues from the nonreducing end of a glycogen chain and attachment of this segment by an α-(1→6) linkage to the chain. This action creates a branch on the glycogen chain that can be extended by the action of glycogen synthase.

6. (a)

$$
\begin{array}{ccc}
\text{COO}^\ominus & & \text{COO}^\ominus \\
| & \text{-GTP} \quad \text{GDP} & | \\
\text{C}=\text{O} & \xrightarrow{\text{PEP carboxykinase}} & \text{C}-\text{OPO}_3^{2\ominus} \\
| & & \| \\
\text{CH}_2 & \text{CO}_2 & \text{CH}_2 \\
| & & \\
\text{COO}^\ominus & &
\end{array}
$$

oxaloacetate   phosphoenolpyruvate (PEP)

(b) Prolonged release of glucagon causes continued elevation of cAMP, which triggers increased transcription of the PEP carboxykinase gene in liver, resulting in increased synthesis of PEP carboxykinase. The rate of gluconeogenesis is increased as a consequence of increased enzyme concentration.

7. (a) The Cori cycle is a combination of glycolysis that occurs in muscle and other peripheral tissues and gluconeogenesis that occurs in liver. Lactate produced by glycolysis in peripheral tissues is carried to the liver via the blood where it is converted to glucose by gluconeogenesis.

(b) The function of the cycle is to allow continued energy production in peripheral tissues via glycolysis. The glucose produced in the liver by gluconeogenesis is transported back to the peripheral tissues via the blood for continued glycolytic metabolism. The Cori cycle helps maintain the level of glucose in the blood.

8. (a) The pentose phosphate pathway is most active in mammary glands, liver, adrenal glands, and adipose tissue.

(b) The enzymes of the pentose phosphate pathway are located in the cytosol.

(c) The principal functions of the pathway are to produce NADPH, required for reductive biosynthesis, and ribose 5–phosphate, which is required for biosynthesis of nucleotides (and, hence DNA and RNA) and related derivatives.

9. (a) 6-Phosphogluconate dehydrogenase catalyzes the oxidative decarboxylation of 6-phosphogluconate to form CO_2, NADPH, and ribulose 5-phosphate.

(b) This reaction occurs in the pentose phosphate pathway and is the last step in the oxidative stage of the pathway.

10. (a) xylulose 5-phosphate + ribose 5-phosphate

(b) xylulose 5-phosphate + erythrose 4-phosphate

(c) sedoheptulose 7-phosphate + glyceraldehyde 3-phosphate

(Note that these are reversals of reactions of the pentose phosphate pathway. Such reversals are possible since all reactions in the nonoxidative stage of the pathway are near equilibrium reactions.)

D. Additional Problems

1. (a) From the data in Table 12.1, add the standard free energy changes, $\Delta G^{0\prime}$, for the reverse of glycolysis reactions 2, 4, 5, 6, 7, 8 and 9 as well as the standard free energy changes for the four reactions unique to gluconeogenesis. Remember to multiply by two the $\Delta G^{0\prime}$ values of those reactions in the triose stage and to change the signs of the $\Delta G^{0\prime}$ values in Table 12.1 for the reversed glycolysis reactions.

$$\Delta G^{0\prime} = [-1.67 + (-24.0) + (-7.66) + (2 \times -6.28) + (2 \times 18.8) + (2 \times -4.44) + (2 \times -1.84) + (2 \times 2.1) + (-16.7) + (-13.8)] \text{ kJ/mol}$$
$$= [-88.95 + 41.8] \text{ kJ/mol}$$
$$= -47.2 \text{ kJ/mol}.$$

   (b) The process is spontaneous as indicated by the negative value of the standard free energy change.

2. The process would not be spontaneous. You determined in problem 12.D.1. that the $\Delta G^{0\prime}$ for glycolysis was -72.8 kJ/mol. If all glycolysis reactions were reversed, the $\Delta G^{0\prime}$ for the process would be $+72.8$ kJ/mol. This process, with its large positive value of $\Delta G^{0\prime}$, would not be spontaneous. Bypassing the three metabolically irreversible reactions of glycolysis with reactions that, overall, have a reasonably large negative change in standard free energy allows gluconeogenesis to be spontaneous.

3. (a)

   γ — Glu — Cys — Gly
              |
              S
              |
              S
              |
   γ — Glu — Cys — Gly

   NADPH, H⊕   glutathione reductase   NADP⊕   →   2γ — Glu — Cys — Gly
                                                                      |
                                                                      SH

   (b) Glucose 6-phosphate dehydrogenase is the enzyme that catalyzes the first reaction of the pentose phosphate pathway.

   (c) Glutathione is required for proper maintenance of red blood cell membranes, as described in the question. Replenishing glutathione from the dimer form requires NADPH, which is produced by the pentose phosphate pathway. A deficiency of glucose 6-phosphate dehydrogenase could lead to insufficient NADPH production that would also limit the amount of glutathione available for red blood cell maintenance.

4. One effect of caffeine is the inhibition of cAMP phosphodiesterase, which prolongs the epinephrine-producing effect of cAMP. Epinephrine (through an increase of cAMP) leads to glycogen degradation to provide glucose for energy. Glycogen phosphorylase *a* is the active form of the enzyme required for glycogen degradation as indicated in text Figure 14.7. Inhibition of this enzyme by caffeine should prevent glycogen degradation. It seems, therefore, that the two actions are opposed to each other.

5. (a) In a unit of the glycogen chain that includes a branch, there would be 16 residues (eight along the main chain and eight in the branch). Dividing the total residues (6000) by 16 per unit gives 375. So, there are about 375 ends.

   (b) Glycogen molecules are synthesized with the reducing end of a short segment attached to the protein glycogenin. Extension of chains and branches is toward the nonreducing ends, since carbon 1 of the residue being attached is connected to carbon 4 of the residue of the existing chain. Therefore, all ends of the glycogen molecule are nonreducing ends.

CHAPTER 15

A. True–False

1. <u>True.</u>

2. <u>True.</u> Cytochromes *a* and a_3 are part of this complex (Complex IV).

3. <u>False.</u> Q (ubiquinol) is a lipid.

4. <u>False.</u> Iron-sulfur clusters are found only in Complexes I, II, and III.

5. <u>False.</u>   Since the two processes are coupled, the inhibition of one stops the other.

6. <u>True.</u>   See section 15.8 of the text and text Figure 15.13.

7. <u>True.</u>   2,4-Dinitrophenol is an uncoupler that disconnects ATP production from electron transport, thus allowing faster oxygen uptake.

8. <u>False.</u>   Oxidation of one mole of NADH produces about 2.5 moles of ATP, but one mole of succinate produces only about 1.5 moles of ATP.

9. <u>True.</u>   According to the chemiosmotic theory, a proton concentration gradient is created by the transfer of protons from the matrix to the intermembrane space of the mitochondrion as electrons are transferred by the electron transport system.

10. <u>True.</u>   In mitochondria, the reverse reaction (formation of ATP from ADP and P_i) occurs, but the F_1-ATPase activity was first detected in knobs isolated from mitochondria.

B. Short Answers

1. cytochromes

2. electron-transporting complexes (respiratory complexes)

3. four; cofactors

4. crossover point

5. Q (ubiquinone)

6. cyanide; carbon monoxide (other inhibitors are also known)

7. Complex II; succinate dehydrogenase

8. stalk and knob structures of the inner mitochondrial membrane (Complex V, ATP synthase)

9. Peter Mitchell; chemiosmotic theory

10. P:O ratio

C. Problems

1. $$2\,NADH + 7\,H^{\oplus} + 5\,ADP + 5\,P_i + O_2 \rightarrow 2\,NAD^{\oplus} + 5\,ATP + 7\,H_2O$$
   (Note: This is based on the ionic states and conversions shown in text Figure 11.6 and in text Table 11.1.)

2. $$2\,succinate + 3\,ADP + 3\,P_i + 3\,H^{\oplus} + O_2 \rightarrow 3\,ATP + 2\,fumarate + 5\,H_2O$$
   (Note: This is based on the ionic states and conversions shown in text Figure 11.6 and in text Table 11.1.)

3. Succinate transfers its electrons to Q (ubiquinone) with no contribution to the formation of the proton concentration gradient. Because the electrons from succinate enter the electron transport chain at a point that bypasses Complex I, the free energy that becomes available when a substrate enters at Complex I is not obtained. Only the free energy obtained from Complexes III and IV contributes to the proton concentration gradient that drives the formation of ATP. (This is consistent with the requirement of a stronger oxidizing agent than $NAD^{\oplus}$ to oxidize

$$—CH_2CH_2—\ to\ —\overset{\textstyle H}{\underset{\textstyle H}{C}}=\overset{\textstyle }{C}—.)$$

4. Use the standard reduction potentials from text Table 11.4 as follows:

   $$NAD^{\oplus} + 2\,H^{\oplus} + 2\,e^{\ominus} \rightarrow NADH + H^{\oplus} \qquad E^{0\prime} = -0.32\ V$$
   $$FMN + 2\,H^{\oplus} + 2\,e^{\ominus} \rightarrow FMNH_2 \qquad\qquad\ E^{0\prime} = -0.30\ V$$

   The equation for the oxidation of NADH and the cell potential is:

   $$NADH + H^{\oplus} + FMN \rightarrow NAD^{\oplus} + FMNH_2 \quad \Delta E^{0\prime} = +0.02\ V$$

   From this, the standard free energy change can be calculated from

   $$\Delta G^{0\prime} = -nF\Delta E^{0\prime},\ where\ n = 2;\ F = 96.48\ kJ\ V^{-1}\ mol^{-1};$$

$$\Delta G^{0\prime} = -(2)(96.48 \text{ kJ V}^{-1} \text{ mol}^{-1})(0.02 \text{ V})$$
$$= -3.86 \text{ kJ/mol} = -4 \text{ kJ/mol}.$$

5. A respiratory chain inhibitor stops both electron transfer and oxidative phosphorylation, since there is no production of a proton concentration gradient. The inhibition site of antimycin A is between cytochrome b_{560} and $\bullet Q^{\ominus}$ or QH_2. Therefore, (a) oxygen is not consumed, (b) no ATP is formed, (c) neither $NAD^{\oplus}$ nor QH_2 is regenerated, and (d) there is no heat production.

6. 2,4-Dinitrophenol uncouples the process of electron transport from the synthesis of ATP. Therefore, (a) oxygen consumption is increased, since electron transfer can occur more rapidly, (b) no ATP is formed, (c) $NAD^{\oplus}$ is regenerated, and (d) heat production is increased, since the energy normally used for ATP formation is released as heat.

7. Text Table 15.3 gives the standard free energy changes, with NADH as the substrate, for electron transfer involving each of the four complexes. The standard free energy change for the entire process is:

$$\Delta G^{0\prime}\text{Total} = [(-70) + (-2) + (-37) + (-110)] \text{ kJ/mol} = -219 \text{ kJ/mol}.$$

This is essentially the same value as determined in text Section 11.9.B.

8. The oxidation of one mole of NADH by the electron transport chain produces 2.5 moles of ATP. Each mole of ATP requires 30 kJ for its formation (the same amount of energy released when ATP is hydrolyzed). Thus, the percent of energy recovered is:

$$\% \text{ energy recovered} = \frac{(2.5 \text{ mol ATP})(30 \text{ kJ/mol})(100)}{[220 + (2.5 \times 30)] \text{ kJ/mol}} = 25.42 = 25\%$$

9. 2,4-Dinitrophenol dissipates the proton concentration gradient. It can release its proton in the basic mitochondrial matrix. The anionic form is sufficiently hydrophobic that it can diffuse through the inner membrane to the cytosol, pick up another proton, move back to release the proton into the matrix, and repeat the process. Thus, the matrix becomes less basic than in the absence of 2,4-dinitrophenol, and the proton concentration gradient between the matrix and the intermembrane space is diminished or lost.

10. A continuing supply of ADP and P_i from the cytosol is required to make ATP. Synthesized ATP must be transported from mitochondria for use in other areas of the cell. Carrier proteins in the inner mitochondrial membrane transport the nucleotides into and out of mitochondria.

D. Additional Problems

1. (a) Because 2,4-dinitrophenol is an uncoupling agent, electron transport continues but little or no ATP is formed. Lowered ATP levels signal the body to mobilize fat stores for energy (ATP) production. (b) Weakness, collapse, and death may occur, depending on the severity of ATP depletion. Oxygen consumption is increased as the burden of ATP formation is removed by uncoupling of the electron transport system. Energy normally used to drive ATP formation is released as heat, elevating body temperature and resulting in profuse sweating in response to this condition.

2. (a) $O_2 + 4 e^{\ominus} + 4 H^{\oplus} \rightarrow 2 H_2O$

   (b) As indicated in text Figure 15.12, two protons are translocated from the mitochondrial matrix to the intermembrane space for each pair of electrons transferred, which results in the reduction of one atom of oxygen to water. The formation of water, the product of oxygen reduction, requires two protons, which are used from those in the mitochondrial matrix. The net result is that the concentration of protons in the intermembrane space is greater by four than that in the mitochondrial matrix.

3. You determined in problem C.4. that the change in standard free energy for the transfer of electrons from NADH to the FMN of Complex I is –4 kJ/mol. Formation of one ATP requires 30 kJ/mol. However, for transfer of electrons to Q,

$$NADH + H^{\oplus} + Q \rightarrow QH_2 + NAD^{\oplus}, \qquad E^{0\prime} = +0.32 \text{ V} + 0.04 \text{ V} = +0.36 \text{ V}$$
$$\Delta G^{0\prime} = -nF\Delta E^{0\prime} = -(2)(96.48 \text{ kJ V}^{-1} \text{ mol}^{-1})(0.36 \text{ V})$$
$$= -69.47 = -69 \text{ kJ/mol}.$$

This is more than enough energy for the formation of one ATP.

4. Thermogenesis is the term for heat generation. Brown adipose tissue (brown fat) is rich in mitochondria. In these mitochondria, long-chain fatty acids, released in response to norepinephrine, uncouple oxidative phosphorylation from electron transport. The energy thus available is released as heat to maintain body temperature.

5. Since neither NADH nor succinate is oxidized, neither Complex I nor Complex II is present. Complex IV catalyzes the transfer of electrons from reduced cytochrome *c* to oxygen. Therefore, Complex IV is present, but there is no evidence for the presence or absence of Complex III.

CHAPTER 16

A. True–False

1. <u>True.</u>   These organisms are dependent, in the dark, on mitochondrial respiration for energy.

2. <u>False.</u>   Although the dark reactions do not depend on light to occur, both the light and dark reactions of photosynthesis occur simultaneously.

3. <u>False.</u>   Oxygen is produced by the light-dependent reactions.

4. <u>False.</u>   Labeling experiments have shown that CO_2 taken in by the plant is fixed as carbohydrate and that oxygen is produced from the oxidation of water. (See the equations in Section 16.1 of your text.)

5. <u>True.</u>   The accessory pigments present in photosynthetic membranes complement the absorptions of chlorophyll *a* and chlorophyll *b* so that light across the visible spectrum is absorbed and used.

6. <u>False.</u>   Green and purple bacteria use bacteriochlorophyll *a* and bacteriochlorophyll *b* for this function.

7. <u>False.</u>   Protons are pumped into the lumen, the aqueous space enclosed by the thylakoid membrane.

8. <u>True.</u>   Plants contain two units, photosystem I and photosystem II, which have slightly different absorption maxima and electron-transport chains.

9. <u>True.</u>   For this reason, the cyclic electron transport sequence operates to form ATP without the simultaneous formation of NADPH. See text Section 16.5.

10. <u>False.</u>   Sucrose is formed in the cytosol, but starch is formed in chloroplasts and amyloplasts.

B. Short Answers

1. light; O_2; ATP; NADPH; dark; carbohydrates

2. chloroplasts

3. thylakoid; grana

4. light-harvesting complexes

5. photophosphorylation

6. reaction center; special pair

7. thylakoid membrane

8. ferredoxin

9. oxygen-evolving complex

10. 3-phosphoglycerate

11. NADPH

12. ribulose-1,5-*bis*phosphate carboxylase; Rubisco

C. Problems

1. In the light reactions, solar energy is used to oxidize water to produce oxygen. Protons derived from water are used in the chemiosmotic synthesis of ATP from ADP and P_i, and electrons from water reduce $NADP^{\oplus}$ to NADPH. In the dark reactions, ATP and NADPH provide the energy and reducing power, respectively, to convert CO_2 to carbohydrates.

2. Sucrose is made from four moles of triose phosphate (shown below as G 3-P), each of which requires three moles of CO_2. Two NADPH are required to reduce each mole of CO_2, and two electrons are transferred from each mole of water to reduce $NADP^\oplus$ to NADPH. Therefore,

moles H_2O = (H_2O/2 $e^\ominus$)(2 $e^\ominus$/NADPH)(2NADPH/CO_2) (3 CO_2/G 3–P)(4 G 3–P/sucrose) = 24 moles H_2O/mole sucrose.

3. Since $E = h\nu$, and $\nu = c/\lambda$, then $E = h\, c/\lambda$. The energy of a mole of photons of 400 nm light is:

   $E = [6.626 \times 10^{-34}$ J s$][(3.00 \times 10^{17}$ nm/s$)/400$ nm$]$

   $= 4.97 \times 10^{-19}$ J/photon $\times 6.02 \times 10^{23}$ photons/mol $= 2.99 \times 10^5$ J/mol $= 299$ kJ/mol.

   The energy of a mole of photons of 700 nm light is:

   $E = [6.626 \times 10^{-34}$ J s$][(3.00 \times 10^{17}$ nm/s$)/700$ nm$]$

   $= 2.840 \times 10^{-19}$ J/photon $\times 6.02 \times 10^{23}$ photons/mol $= 1.71 \times 10^5$ J/mol $= 171$ kJ/mol.

   The photon of light of shorter wavelength, 400 nm, does provide a greater amount of energy.

4. The carotenoids are accessory pigments or antenna pigments that, like chlorophyll, are capable of absorbing light. They absorb light of wavelengths not absorbed maximally by chlorophyll and transfer the absorbed energy to adjacent pigments in the photosystem. The energy is ultimately transferred to the special pair chlorophylls in the reaction center.

5. Similarities include:

   a. location within a cellular organelle

   b. membrane-bound components

   c. transfer of electrons via system components

   d. production of a proton concentration gradient

   e. formation of ATP

   f. an ATP synthase with a stalk and knob structure

6. If 8 photons produce one O_2, and one photon excites only one in 2400 chlorophyll molecules, (8 photons/O_2) × (O_2/2400 chlorophyll) = 1 photon per 300 chlorophyll. This indicates that 300 chlorophyll molecules constitute a functional cluster, which utilizes one photon. It must also indicate that eight such clusters are necessary to produce one O_2 molecule.

7. It takes two molecules of water to produce one oxygen molecule, two ATP, and two NADPH molecules. It would take six times this process to provide 12 ATP and 12 NADPH. Fixation of six CO_2 as one hexose requires 18 ATP and 12 NADPH. Thus, it appears there is a deficiency of 6 ATP. More ATP is formed without concomitant NADPH formation by cyclic electron transport, as described in text Section 16.5. High NADPH concentrations favor cyclic electron flow.

8. PSI is located in the stromal lamella and PSII is located in the granal lamellae. This difference in location prevents direct transfer of excitation energy between the two photosystems, which would short circuit energy transfer. They are linked by specific electron carriers and work in series.

9. This study supported Mitchell's theory regarding ATP formation via chemiosmosis. The artificially-generated proton concentration gradient drives the formation of ATP. It suggested that the same kind of process occurs when ATP is formed during photophosphorylation (i.e., illumination causes accumulation of protons in the lumen.)

10. (a) Six molecules of CO_2 are required.

    (b) Each of the six CO_2 molecules combines with a ribulose-1,5-*bis*phosphate molecule, forming a total of 12 3-phosphoglycerate molecules. Of these 12, two are used to produce one hexose and 10 go through reactions of the pentose phosphate pathway to regenerate six ribulose-1,5-bisphosphate molecules.

D. Additional Problems

1. The special pair chlorophyll molecules and those in the antenna complex are structurally identical. However, the special pair chlorophyll molecules are in a different physical environment that causes them to act as an electron sink with respect to the chlorophyll molecules of the antenna complex. Figure 16.2 indicates schematically how energy excites ground state antenna complex chlorophylls to their excited states and how this energy is ultimately transferred to reaction center chlorophylls because the excited states of the reaction center chlorophylls are lower than those of the antenna complex chlorophylls.

2. Figure 16.1 shows that the maximum yield of ATP occurs at an external pH of about 8.3-8.4. The yield is about 0.2 ATP per chlorophyll or one ATP per 5 chlorophyll molecules. Emerson and Arnold found that one photon excited

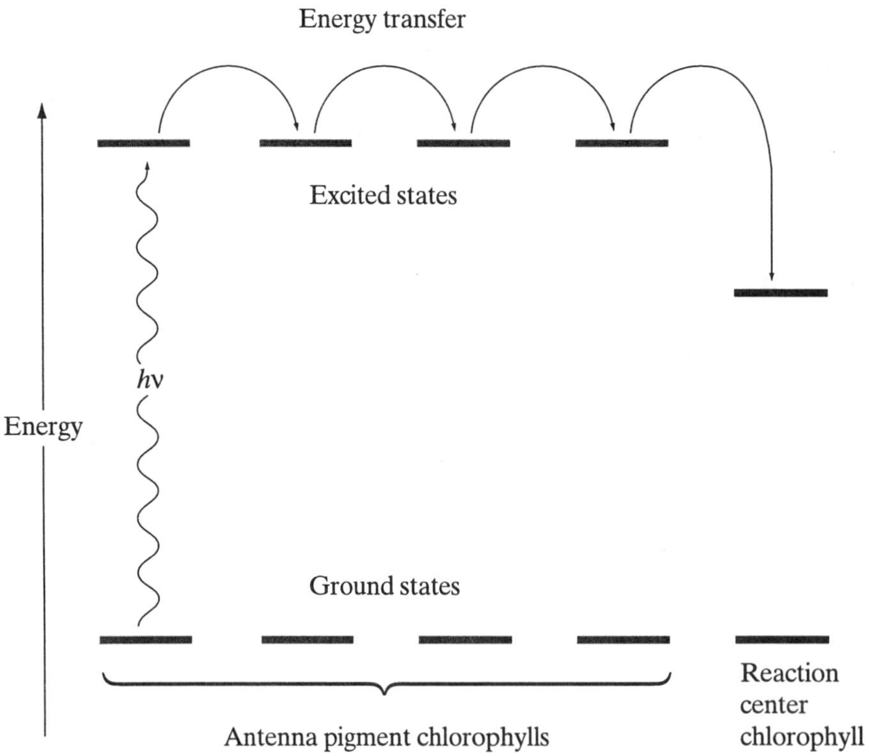

Figure 16.2 Energy Transfer from Antenna Chlorophylls to Reaction Center Chlorophylls.

300 chlorophyll molecules. It takes 8 photons to form one O_2 and 2 ATP. Therefore, 8×300 or 2400 chlorophyll molecules produce 2 ATP, which is one ATP per 1200 chlorophyll molecules. The much greater ATP yield in the Jagendorf and Uribe study is due to the energy buildup (from the formation of a proton concentration gradient) during the period in which the chloroplasts were exposed to light in the absence of ADP.

3. (a) Your text indicates that, normally, the rate of carboxylation in healthy plants is three to four times that of the oxygenation reaction (photorespiration).

(b) In some plants, photosynthesis, on a hot bright day, will deplete the level of CO_2 at the chloroplast and raise the level of O_2 such that the rate of photorespiration may approach that of photosynthesis. This is a major limiting factor in the growth of many plants. (Note: The oxygenase activity of Rubisco increases with temperature more rapidly than does the carboxylase activity.)

(c) In the C4 pathway, C4 acids are decarboxylated in bundle sheath cells, which results in increased CO_2 concentration inside the cells and creates a high ratio of CO_2 to O_2. This minimizes oxygenase activity of Rubisco (photorespiration) and does not limit growth of the plant due to loss of CO_2 that occurs in plants lacking the C4 pathway. (See text Figure 16.19.)

4. The energy for one hour is:

$$7 \text{ J (cm}^2 \text{ s)}^{-1} \times 10 \text{ cm}^2 \times 3600 \text{ s hr}^{-1} = 252 \text{ kJ/hr.}$$

The potential for glucose formation is:

$$252 \text{ kJ hr}^{-1} \times 1 \text{ mol}/2870 \text{ kJ} \times 180. \text{ g/mol} = 15.8 \text{ g.}$$

If only 5.53 grams form, the percent efficiency = $5.53 \text{ g} \times 100/15.8 \text{ g} = 35\%$.

5. $\Delta G^{0\prime}$ for ATP + $H_2O \rightarrow$ ADP + P_i is $-30.$ kJ/mol. ΔG for ATP synthesis is equal to $\Delta G^{0\prime}$ plus the free energy change due to concentrations other than unity. That is:

$$\Delta G = \Delta G^{0\prime} + RT \ln([ATP]/[ADP][P_i])$$

But ΔG for the reaction must be equal to three times the free energy change due to transfer of one proton:

$$\Delta G = 3(2.3 \text{ RT } \Delta pH + ZF\Delta\Psi).$$

But $\Delta\Psi$ is zero due to the transfer of $Ca^{2\oplus}$ and $Cl^{\ominus}$ ions in and out of the thylakoid membrane as protons are pumped through the thylakoid membrane. Therefore, the term $ZF\Delta\Psi$ becomes zero. Then,

$$\Delta G^{0\prime} + RT \ln Q = 3(2.3 \, RT \, \Delta pH),$$

where Q is the ratio of concentrations, $[ATP]/[ADP][P_i]$. Rearranging gives:

$$\Delta pH = \frac{\Delta G^{0\prime} + RT + \ln Q}{3(2.3RT)}.$$

$$\Delta pH = \frac{30. \, \text{kJ/mol} + (8.315 \, \text{J mol}^{-1} \, \text{K}^{-1})(298 \, \text{K})(\ln 1000)}{6.9 \, (8.315 \, \text{J mol}^{-1} \, \text{K}^{-1})(298 \, \text{K})} = \frac{(30. + 17.1) \, \text{kJ mol}^{-1}}{17.1 \, \text{kJ mol}^{-1}} = 2.75 \text{ or } 2.8 \text{ pH units.}$$

This shows that a pH gradient in chloroplasts of about 3 pH units exists across the thylakoid membrane.

6. (a) Since 3-phosphoglycerate is a 3-carbon molecule, the acceptor might be a 2-carbon molecule.

(b) The formation of two molecules of 3-phosphoglycerate for each CO_2 suggested that a 5-carbon molecule might be the acceptor and the resulting 6-carbon intermediate was immediately cleaved to two molecules of 3-phosphoglycerate.

(c) The label is found in the carboxyl group of 3-phosphoglycerate. (See text Figure 16.14.)

CHAPTER 17

A. True–False

1. <u>False.</u>   Carnitine is involved in the transport of fatty acids from the cytosol into mitochondria.

2. <u>True.</u>   Fatty acids with odd-numbered carbon chains are rare.

3. <u>True.</u>   Each two-carbon unit becomes an acetyl CoA.

4. <u>False.</u>   Eicosanoic acid contains 20 carbons and yields 10 acetyl CoA from β-oxidation. Only 9 NADH and 9 QH_2 are formed, since the ninth round of reactions yields two molecules of acetyl CoA as the four-carbon substrate undergoes thiolysis.

5. <u>True.</u>   One ATP is required for the activation. It is converted to AMP and PP_i. Because PP_i is subsequently hydrolyzed to P_i, the cost of fatty acid activation is two high-energy phosphate bonds.

6. <u>False.</u>   β-Oxidation occurs in mitochondria, but fatty acid synthesis occurs in the cytosol.

7. <u>True.</u>

8. <u>False.</u>   The pentose phosphate pathway furnishes about half of the NADPH required for fatty acid synthesis and the rest is furnished by the citrate transport system.

9. <u>True.</u>   The side chain carbon atoms also are derived from acetyl CoA, which is transported from the mitochondria into the cytosol by the citrate transport system.

10. <u>True.</u>   LDL-derived cholesterol can accumulate on the inner arterial walls.

11. <u>True.</u>   Animals cannot create double bonds in fatty acyl CoA carbon chains beyond carbon 9. Linoleate $(18:2 \, \Delta^{9,12})$ is a diet requirement for most animals.

B. Short Answers

1. acetyl CoA; NADH; QH_2

2. thiolase

3. β-hydroxybutyrate; acetoacetate; acetone

4. mitochondria; peroxisomes

5. malonyl CoA

6. desaturases

7. local regulators; arachidonate

8. leukotrienes

9. Lovastatin

10. geranyl pyrophosphate; farnesyl pyrophosphate; squalene

C. Problems

1. Stearoyl CoA + 8 Q + 8 NAD$^\oplus$ + 8 CoASH + 8 H$_2$O → 9 Acetyl CoA + 8 QH$_2$ + 8 NADH + 8 H$^\oplus$

2. In the activation of a fatty acid, one ATP is converted to AMP and PP$_i$. Pyrophosphate is then hydrolyzed by the action of pyrophosphatase, helping to drive the activation reaction to completion. Therefore, two phosphoanhydride bonds are consumed.

3. The unsaturated intermediate in the β-oxidation pathway has a *trans* double bond between carbons 2 and 3. Natural unsaturated fatty acids usually have *cis* double bonds that are usually not in the same position as the *trans* bonds produced during β-oxidation. Text Figure 17.11 shows the steps necessary for the oxidation of linoleoyl CoA.

4. These ketone bodies are formed when acetyl CoA is present in excess of the amount that can be oxidized by the citric acid cycle. These water-soluble molecules are routed to tissues such as the heart and kidney to supply part of their energy needs. The ketone bodies are more readily transported in the blood than fatty acids.

5. Bicarbonate adds to acetyl CoA to form malonyl CoA, in which the free carboxylate group was supplied by bicarbonate. When malonyl CoA adds either to an acetyl unit or the growing fatty acid, the free carboxylate group of malonyl CoA is released as CO$_2$. Therefore, no label is incorporated into palmitate.

6. When cholestyramine reaches the intestinal tract, the anionic bile salts are attracted to the positively-charged resin and are eliminated from the body. Consequently, LDL is removed from the blood by the liver to release cholesterol, the precursor required for the synthesis of bile salts, to replace those excreted.

7. The ATP yield per carbon atom increases slightly with the length of the carbon chain. Your text indicated that palmitate yields 106 ATP (Section 17.2.D). For stearate, approximately 90 ATP are generated from 9 acetyl CoA, 12 from 8 QH$_2$, and 20 from 8 NADH, for a total of 122 ATP. Subtracting 2 ATP for activation, the net yield is 120 ATP. The normalized ATP yield for glucose is 32 ATP × 18/6 = 96 ATP. The percentage increase that stearate yields with respect to that of glucose is:

$$(120 - 96)(100)/96 = 25\%.$$

Therefore, the percentage increase is about one percent higher for stearate than it is for palmitate.

8. (a) The citrate transport system operates to export acetyl CoA from mitochondria to the cytosol and to produce NADPH in the cytosol. Both acetyl CoA and NADPH are required for cytosolic fatty acid synthesis.

   (b) The system requires energy. Two ATP are required for each acetyl CoA exported to the cytosol.

9. (a) Pyruvate + HCO$_3^\ominus$ + ATP $\xrightarrow{\text{pyruvate carboxylase}}$ Oxaloacetate + P$_i$ + ADP

   (b) Citrate + ATP + CoASH $\xrightarrow{\text{citrate lyase}}$ Oxaloacetate + Acetyl CoA + ADP + P$_i$

   (Note: Reaction 9. (a) occurs in the mitochondrial matrix and reaction 9. (b) occurs in the cytosol.)

10. The key regulatory enzyme in fatty acid synthesis is acetyl-CoA carboxylase, which catalyzes the carboxylation of acetyl CoA to form malonyl CoA. The reaction is metabolically irreversible. Citrate activates the enzyme, which is inactivated by phosphorylation. Glucagon stimulates this phosphorylation reaction.

11. Sphingomyelin (see text Figure 17.26) is subject to base-catalyzed hydrolysis and cholesterol is not. Reflux samples of both solutions with aqueous NaOH. Fatty acid salts, that will dissolve in the aqueous phase, will be produced from sphingomyelin. Acidification of the aqueous phase should cause precipitation of the free fatty acids. Cholesterol does not react.

D. Additional Problems

1. Lack of carbohydrates forces the body to rely on fatty acid oxidation for energy. Supplies of pyruvate and, consequently, oxaloacetate would be inadequate for energy needs due to the lack of carbohydrates. With insufficient oxaloacetate, little acetyl CoA could be consumed in the citric acid cycle. Abnormal quantities of the acetyl CoA would be converted to acetoacetate and β-hydroxybutyrate. Large amounts of these ketone bodies would lead to a drastic lowering of blood pH that could be life-threatening if not checked.

2. Eventually, the supply of CoASH would be depleted by unchecked fatty acid oxidation and energy production would be blocked. Under these conditions, acetoacetate is produced from the accumulating acetyl CoA, a process that releases two CoASH molecules per acetoacetate formed.

3. The LDL-receptor proteins are involved in the removal of LDL from the blood. In individuals with FH, there is essentially no mechanism for removing cholesterol, as LDL, from the blood. Cholesterol derived from LDL also serves to diminish cholesterol synthesis. This control would also be lacking.

4. The protein avidin strongly binds biotin, the prosthetic group of the enzyme acetyl CoA carboxylase. This enzyme catalyzes the carboxylation of acetyl CoA to form malonyl CoA. Since each two-carbon unit added to the growing fatty acid chain is donated by malonyl CoA, prevention of its formation blocks fatty acid synthesis.

5. When fatty acids with odd-number carbon chains are oxidized, the last round of β-oxidation produces propionyl CoA that is converted to succinyl CoA, which can be converted to oxaloacetate. Since oxaloacetate is a substrate for gluconeogenesis, the carbons of the original propionyl CoA molecule are converted to glucose. This is a minor exception to the general rule because fatty acids with an odd-number of carbons are rare in nature.

CHAPTER 18

A. True–False

1. <u>False.</u> Approximately one half of the 20 amino acids used for protein synthesis in mammals must be obtained in the diet. The number varies among species and with the stage of development.

2. <u>False.</u> While a portion of the amino acids from dietary proteins is used for cellular protein synthesis, some are deaminated and their carbon skeletons used for other biosynthetic purposes, or to generate energy.

3. <u>True.</u> L-Glutamate is a substrate that donates an amino group to some α-keto acid acceptor. α-Ketoglutarate is a substrate that serves as an acceptor of an amino group from some amino acid. The transaminases differ in their specificities for particular amino acids and α-keto acids.

4. <u>False.</u> The liver is the site of urea synthesis.

5. <u>True.</u> The process of the reduction of one molecule of N_2 to $2 NH_3$ requires a minimum of 16 ATP.

6. <u>True.</u> This means of carrying $NH_4^{\oplus}$ keeps blood levels of this toxic metabolite low.

7. <u>False.</u> Some amino acids are both glucogenic and ketogenic.

8. <u>False.</u> High levels of arginase are found only in the liver. It catalyzes the hydrolysis of arginine to form ornithine and urea.

9. <u>True.</u> Bicarbonate is replenished by glutamine catabolism in the kidneys. Each molecule of glutamine furnishes two bicarbonate ions.

10. <u>True.</u> Nitric oxide is a messenger molecule with several physiological functions. Nitric oxide helps kill bacteria and tumor cells, but is harmful at high concentrations.

B. Short Answers

1. pyridoxal phosphate; to act as a carrier of amino groups

2. lysine; threonine

3. ketogenic; glucogenic

4. phenylketonuria (PKU)

5. cumulative feedback inhibition

6. $NH_4^{\oplus}$; aspartate; $HCO_3^{\ominus}$

7. carbamoyl phosphate

8. four

9. glucose-alanine cycle

10. the glycine-cleavage system

C. Problems

1. Amino groups are first released from the amino acids as ammonium ions. The remaining carbon chains of the amino acids are then oxidized for energy by separate pathways.

2. Figure 18.1 shows the reaction, catalyzed by alanine transaminase, in which alanine is converted to pyruvate, which can be converted to oxaloacetate, which can ultimately form glucose by gluconeogenesis. (Note that this reaction is essentially the reverse of that shown in text Section 18.4.A. where the synthesis of alanine is illustrated.)

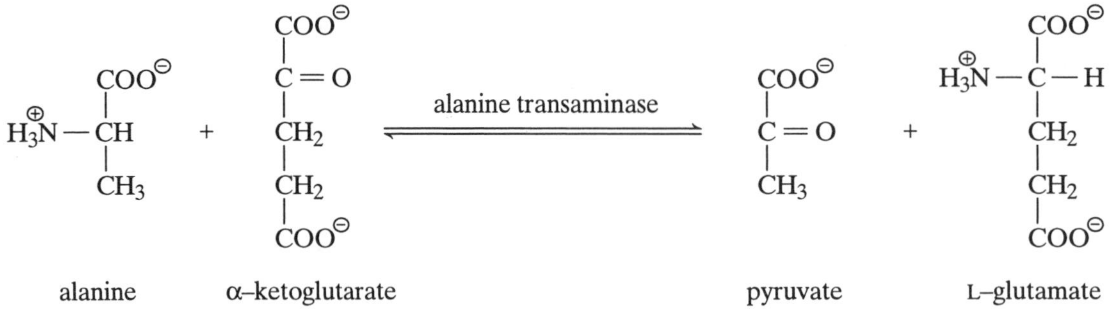

Figure 18.1 Conversion of Alanine to Pyruvate in Liver.

3. The positive value of the standard free energy change, +27 kJ/mol, indicates that, under standard conditions (1 M concentrations), the formation of α-ketoglutarate is not spontaneous. In cells, however, concentrations are not 1 M and, further, the $NH_4^\oplus$ ions produced are constantly being removed and this drives the reaction toward continuous production of α-ketoglutarate.

4. The formation of carbamoyl phosphate requires 2 ATP and releases 2 ADP and P_i as one amino group is brought to the urea cycle. The formation of argininosuccinate from aspartate and citrulline uses one ATP and forms AMP and PP_i. Subsequent hydrolysis of PP_i cleaves a high-energy bond that, together with the previous 3 ATP used, makes a total four high-energy phosphate bonds consumed.

5. The first committed step of the urea cycle, the formation of carbamoyl phosphate, is catalyzed by carbamoyl phosphate synthetase I. *N*-Acetyl glutamate is an activator of this enzyme. The synthesis of this activator increases as glutamate concentrations increase as a result of transamination reactions involving amino acids being metabolized. The rates of reactions of the urea cycle are controlled by concentrations of their substrates.

6. Text Figure 18.10 illustrates the process by which glycine is produced from serine by the action of serine hydroxymethyltransferase, with pyridoxal phosphate as the prosthetic group. Tetrahydrofolate, derived from folate by reduction, is a coenzyme required for the reaction. Therefore, glycine synthesis is dependent both on levels of serine and folate.

7. Figure 18.2 illustrates the cyclization of glutamate γ-semialdehyde to form a Schiff base that is then reduced to form proline.

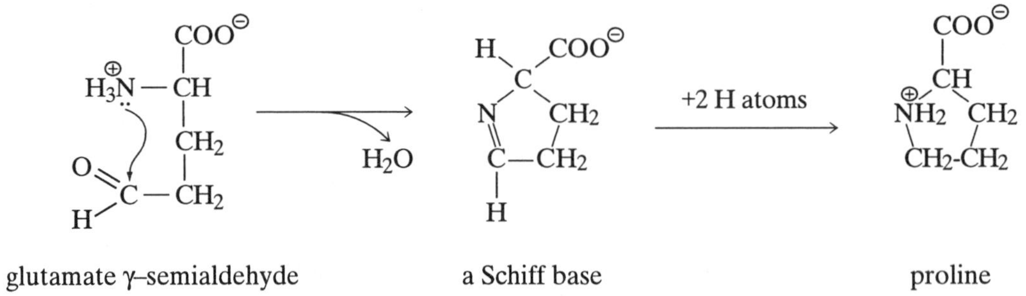

Figure 18.2 Final Steps of Proline Synthesis.

8. The seven compounds are: oxaloacetate, fumarate, pyruvate, acetyl CoA, α-ketoglutarate, succinyl CoA, and acetoacetyl CoA.

9.

Ketogenic	Glucogenic	Glucogenic-Ketogenic
leucine	alanine	phenylalanine
lysine	cysteine	tyrosine
	glycine	isoleucine
	serine	threonine
	asparagine	tryptophan
	aspartate	
	methionine	
	valine	
	arginine	
	glutamate	
	glutamine	
	histidine	
	proline	

10. Synthesis of tyrosine depends on the presence of the essential amino acid phenylalanine. Cysteine synthesis involves homocysteine, which is derived from the essential amino acid methionine.

D. Additional Problems

1. Urease catalyzes the conversion of urea to ammonia and carbon dioxide. The following equation shows that three moles of gas are produced per mole of urea.

$$\underset{\text{H}_2\text{N}-\overset{\displaystyle\overset{\text{O}}{\|}}{\text{C}}-\text{NH}_2}{} + \text{H}_2\text{O} \xrightarrow{\text{urease}} 2\,\text{NH}_3 + \text{CO}_2$$

Mol urea = 2.4 mL gas $\times$ (1 mol gas/22.4 $\times$ 10^3 mL gas) $\times$ (1 mol urea/3 mol gas products) = 3.6 $\times$ 10^{-5} mol urea.

2. The alanine-glucose cycle operates to remove both pyruvate and $\text{NH}_4^{\oplus}$ from muscle, in the form of alanine, and to deliver it to the liver, where a reversal of the synthetic reaction releases $\text{NH}_4^{\oplus}$ and pyruvate. The $\text{NH}_4^{\oplus}$ is removed by incorporation into urea. Glucose is formed from pyruvate by gluconeogenesis and can be returned to muscle for energy production.

The Cori cycle shuttles lactate, formed by reduction of pyruvate, from muscle to liver, where gluconeogenesis also produces glucose that can be returned to muscle.

The alanine-glucose cycle requires equal amounts of $\text{NH}_4^{\oplus}$ and pyruvate. If production of pyruvate in muscle exceeds $\text{NH}_4^{\oplus}$ present, the Cori cycle operates to cycle the excess pyruvate.

Defects in the urea-cycle enzymes can lead to high [$\text{NH}_4^{\oplus}$].

3. Although the ammonium ion cannot enter the brain, it is in equilibrium with NH_3, which is able to enter the brain. Upon entry, NH_3 becomes protonated to $\text{NH}_4^{\oplus}$. Parts of the brain utilize glycolysis, the citric acid cycle, and related pathways for energy requirements. $\text{NH}_4^{\oplus}$ may cause depletion of α-ketoglutarate and glutamate by reacting with them to form glutamate and glutamine, respectively. With a citric acid cycle intermediate in short supply, energy production would be curtailed. Additionally, glutamate is a neurotransmitter and is a precursor for another neurotransmitter, γ-aminobutyrate. Drastic alteration of neurotransmitter levels as well as reduced levels of energy production may lead to brain damage.

4. Aspartate is the form in which the second nitrogen atom of urea enters the urea cycle. Aspartate is formed by a transamination reaction in which the amino group from some amino acid is transferred to oxaloacetate.

5. (a) The ^{18}O label is found in AMP in the reaction:

Citrulline + Aspartate + ATP $\longrightarrow$ AMP + PP$_i$ + argininosuccinate

(b) This finding suggests the formation of a citrullyl-AMP intermediate that is subsequently cleaved to yield AMP and argininosuccinate.

CHAPTER 19

A. True–False

1. <u>False.</u> The purine fused-ring system contains nine atoms and pyrimidine rings contain six atoms.

2. <u>True.</u>

3. <u>True.</u> The sources of the various atoms of the purine ring system are illustrated in text Figure 19.2.

4. <u>False.</u> The purine ring components are assembled on ribose 5-phosphate. That is, ribose 5-phosphate serves as a foundation upon which the purine ring is built.

5. <u>True.</u> IMP, which contains the purine base hypoxanthine, is converted to AMP by one branch of the pathway and to GMP by the other branch of the pathway.

6. <u>True.</u> See text Figure 19.7.

7. <u>True.</u> See text Figure 19.9.

8. <u>False.</u> Orotate is an intermediate in the pyrimidine biosynthetic pathway, but uridine 5′-monophosphate (UMP) is the end product of the pathway.

9. <u>False.</u> The pyrimidine ring system is completed prior to the attachment of ribose 5-phosphate. The attachment occurs in the reaction between orotate and 5-phosphoribosyl 1-pyrophosphate (PRPP). See text Figure 19.10.

10. <u>False.</u> dTMP is formed by the methylation of dUMP. See text Figure 19.15.

11. <u>False.</u> Degradation of the purines leads to uric acid, which is the end-product in birds and primates, but there are no distinctive excretory products formed from pyrimidine catabolism.

B. Short Answers

1. PRPP (5-phosphoribosyl 1-pyrophosphate)

2. the cytosol

3. the smaller (imidazole) ring (containing five atoms)

4. IMP; XMP; GMP; AMP (See text Figure 19.8.)

5. pyrimidines

6. Uric acid

7. glutamine (See text Figure 19.12.)

8. adenine phosphoribosyltransferase; hypoxanthine-guanine phosphoribosyltransferase

9. purine nucleotide cycle (see text Figure 19.26.)

10. acetyl CoA; succinyl CoA (see text Figure 19.27.)

C. Problems

1. (a) 1 cytosine, 1 ribose

    (b) 1 hypoxanthine, 1 ribose, 1 inorganic phosphate

    (c) 1 guanine, 1 deoxyribose, 3 inorganic phosphates

2. Glutamine furnishes amino groups in steps 1 and 4 of the biosynthesis of IMP from PRPP, and also in the formation of GMP from XMP. The nitrogens of the donated groups become nitrogen atoms 9 and 3, respectively, of the purine ring, and the 2-amino group of GMP. (See text Figures 19.5 and 19.7.)

3. Text Figure 19.3 shows that one ATP is used to form PRPP, but two high-energy bonds are consumed since the pyrophosphate group of PRPP is removed in step 1 of the de novo synthesis and is hydrolyzed to 2 P_i. Text Figure 19.5 shows that four other ATP molecules are used (steps 2, 4, 5, 7) and one high-energy bond of each is consumed. Text Figure 19.7 shows that one GTP is required and one of its high-energy bonds is consumed. This gives a total of seven ATP equivalents (high-energy bonds).

4. (a) In such cases, it is likely that the allosteric enzyme does have more than one binding site for the inhibitors. This enzyme has separate sites for the binding of guanine and adenine nucleotides.

(b) It is also likely that complete inhibition is not achieved by the binding of any one single inhibitor. It is more likely that the greatest inhibition is caused by the binding of all of the inhibitors.

5. 2 ATP + glutamine + 2 H_2O + Q + PRPP → 2 ADP + 4 P_i + glutamate + QH_2 + UMP

6. Unlike most enzymes that catalyze carboxylation reactions, which require ATP and the prosthetic group biotin, AIR carboxylase requires neither biotin nor ATP. The mechanism for this carboxylation is shown in text Figure 19.6.

7. Figure 19.1 illustrates the formation of CTP from UTP.

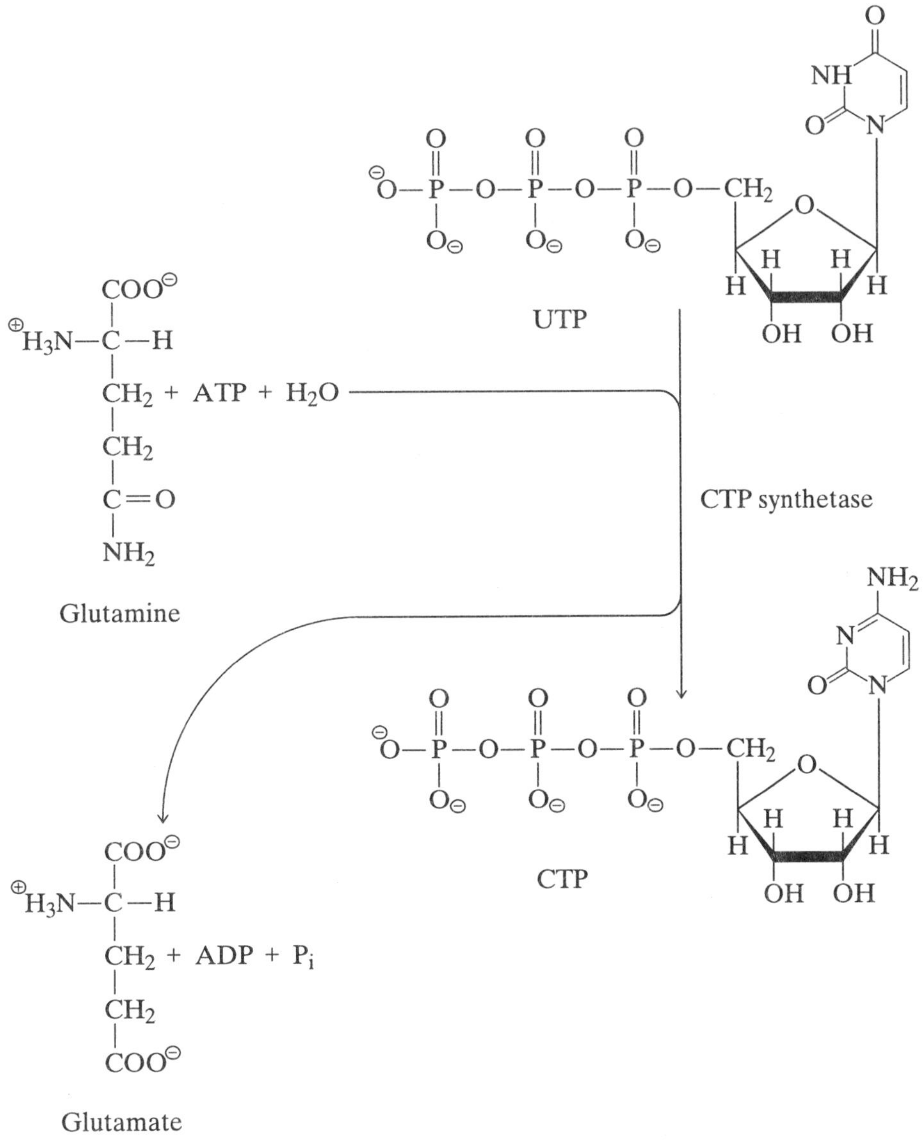

Figure 19.1 CTP Synthesis from UTP.

8. The label will appear in the 5-methyl group of dTMP. (See text Figure 19.15.)

9. Deoxyribonucleotides are synthesized by the reduction of ribonucleotides. An enzyme, ribonucleoside diphosphate reductase, catalyzes the reduction of all four ribonucleoside diphosphates (ADP, GDP, CDP, UDP). Reducing power is supplied by NADPH and is transferred to the ribose moiety by the involvement of FAD, thioredoxin, and sulfhydryl groups of the enzyme.

10. (a) Lesch-Nyhan syndrome results from a hereditary deficiency of hypoxanthine-guanine phosphoribosyltransferase, a salvage enzyme that converts hypoxanthine and guanine to IMP and GMP, respectively. If not so salvaged, free hypoxanthine and guanine are degraded to uric acid.

(b) The disease is generally restricted to males because the gene that codes for the deficient enzyme is located on the X chromosome.

(c) Symptoms include mental retardation, palsy, and self-mutilation.

(d) Since much of the hypoxanthine and guanine normally salvaged is degraded to uric acid, increased de novo synthesis is likely to result from decreased inhibition of glutamine-PRPP amidotransferase by IMP and GMP.

D. Additional Problems

1. Glutamine furnishes N-3 and $HCO_3^{\ominus}$ furnishes C-2 of the pyrimidine ring. (See text Figure 19.9.) The CO_2 eliminated when OMP is converted to UMP was supplied by aspartate. (See text Figure 19.10.) Therefore, both OMP and UMP should contain the ^{14}C-label, but CO_2 should not.

2. In the previous chapter of this study guide, problems 18 II.B.8. and 18 II.C.4. showed that 4 ATP are required for the synthesis of one molecule of urea. Incorporation of one $NH_4^{\oplus}$ into urea then requires 2 ATP. One GTP is required in the reaction of IMP and aspartate to produce adenylosuccinate and GDP and P_i. Therefore, the energy requirement to produce one fumarate by the purine nucleotide cycle, and the elimination of $NH_4^{\oplus}$, is the equivalent of 3 ATP molecules.

3. dUTPase catalyses the hydrolysis of dUTP to dUMP and PP_i. This action prevents dUTP from being incorporated into DNA in place of dTTP. Incorporation of dUTP in any quantity into DNA would interfere with replication and transcription. Cellular death would be the ultimate effect.

4. Increased availability of glucose 6-phosphate, due to the deficiency in glucose 6-phosphatase, stimulates the pentose phosphate pathway in the liver. Increased production of ribose 5-phosphate leads to increased amounts of PRPP, which stimulate purine biosynthesis. The latter leads to higher levels of uric acid and, hence, to gout.

5. (a) In the biosynthetic pathway for IMP, the enzymes of steps 1 and 4, glutamine-PRPP amidotransferase and FGAM synthetase would be inhibited since they both utilize glutamine as a substrate. GMP synthetase, which catalyzes the glutamine-dependent conversion of XMP to GMP, would be inhibited by azaserine, as would carbamoyl phosphate synthetase in the UMP biosynthetic pathway. Other enzymes including CTP synthetase would also be inhibited. (See text Figures 19.5, 19.7, 19.10, and 19.12.)

(b) Affinity labels bind covalently to active site groups, causing irreversible inhibition. Azaserine is known to covalently bind to the sulfhydryl group of a cysteinyl residue in the active sites of several of these glutamine-dependent enzymes. Figure 19.2 illustrates this kind of inhibition.

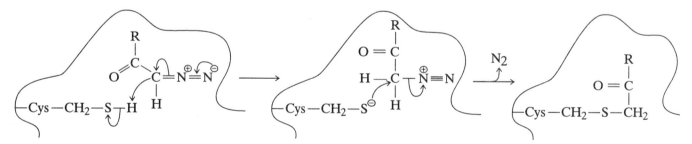

Figure 19.2 Alkylation of an Active-Site Cysteine by a Diazo-antibiotic Such as Azaserine

CHAPTER 20

A. True–False

1. <u>True.</u> Miescher isolated the material in 1869.

2. <u>False.</u> %A = %T and %G = %C, but %(A + T) can be any value (technically) between 0 and 100%. In naturally occurring DNA samples, the %(A + T) ranges from about 30% to about 70%. (See Figure 20.1.)

3. <u>False.</u>   The Watson-Crick model of DNA is an antiparallel, double-stranded, right-handed helix.

4. <u>False.</u> The separated DNA strands absorb more uv radiation because the bases are deshielded.

5. <u>True.</u>   These enzymes are not restricted to cleaving terminal phosphodiester bonds as are exonucleases.

6. <u>True.</u>   While the majority of DNA in eukaryotic cells (that in the nucleus) is not circular, DNA in mitochondria and chloroplasts is circular. In general, bacterial DNA is also circular.

7. <u>False.</u>   DNA is not subject to base-catalyzed hydrolysis as RNA is.

8. <u>True.</u>   See text Section 20.6.A and B.

9. <u>True.</u>   The restriction endonucleases, however, will not catalyze hydrolysis of the DNA strand if the recognition sequence is methylated.

10. <u>False.</u>   Sticky ends are formed when restriction endonucleases catalyze cleavage of the two strands of DNA at different places. This leaves duplex DNA with one strand at each cleavage site longer than the other. The DNA is not denatured.

B. Short Answers

1. B-DNA

2. Z-DNA

3. adenine; guanine

4. melting (thermal denaturation); T_m, the melting temperature

5. restriction endonucleases

6. restriction sites (Many restriction sites are palindromic.)

7. histones; nucleosome

8. hydrogen bonds; charge-charge interactions; hydrophobic effects; stacking interactions (London dispersion forces)

9. topoisomerases

10. messenger RNA (mRNA)

C. Problems

1. The cytosine content is the same (17%) as the guanine content since they form complementary base pairs. Adenine forms complementary base pairs with thymine and the total of adenine + thymine = $100 - 2(17\%) = 66\%$. Therefore, the %adenine = %thymine = 33%.

2. The average rise in a DNA chain (distance between bases) is 0.33 nm. The number of bases will be:

     bases = 1 base pair/0.33 nm × 51 nm × 2 bases/pair = 315 or 320 bases (to two significant figures).

3. The base sequence C-G-C-G-C-G is complementary to itself in the opposite direction.

     5′-C-G-C-G-C-G-3′

     3′-G-C-G-C-G-C-5′

4. Figure 20.1 shows that a T_m of 90°C corresponds to a G + C content of about 51 mol%, which means that the A + T content is 49 mol%, and the adenine is 24.5 mol%.

5. The T_m would be higher than that of the DNA in the previous problem. With a lower percentage of A-T pairs, there would be more G-C pairs, thus the stacking interactions would be stronger and there would be more hydrogen bonds and a higher T_m. From Figure 20.1, the T_m is about 99°C.

6. Some bacteria have both restriction methylases and restriction endonucleases that recognize the same sequences in DNA. The methylases catalyze the addition of methyl groups to these sequences of the cell's own DNA. This serves as a protection and prevents the cell's restriction endonucleases from cleaving its own DNA. Unmethylated foreign DNA is unprotected and is cleaved by the bacterial cell's restriction endonucleases. (Also note that hemimethylated sites of newly-replicated DNA are high-affinity substrates for the methylase but are not recognized by the restriction endonuclease. This assures continued protection of the cell's own DNA.)

7. As complementary bases form pairs, the pairs are between the two chains. The base pairs are stacked above and below each other between the two chains. Cooperative interactions between the base pairs bring the pairs closer together and cause the sugar-phosphate backbone to twist into the helix shape. These stacking interactions are responsible in part for the helix formation.

8. Ribosomal RNA, rRNA, makes up about 80% of all cellular RNA. It is a component of the cytoplasmic particles called ribosomes, which are involved in protein synthesis.

9. Nucleosomes are composed of histones and DNA. A histone octamer, called a core particle, forms from two each of histones H2A, H2B, H3, and H4. About 146 base pairs of DNA wrap around the core particle, with a total of 200 base pairs being involved in the formation of a nucleosome. An H1 histone can also be associated with the linker DNA between nucleosomes. Coiling of the beads-on-a-string assembly of nucleosomes gives rise to a more condensed structure called the 30-nm fiber.

10. *Hae*III and *Sma*I catalyze nonstaggered cuts, producing blunt ends. These enzymes catalyze the cleavage of both strands of duplex DNA at the same location so that no single-stranded segments, or sticky ends, are formed.

D. Additional Problems

1. (a) Since there are 200 base pairs per nucleosome, the total number of nucleosomes in the chromosome would be:
$$(2.4 \times 10^8 \text{ bp})(1 \text{ nucleosome/200 bp}) = 1.2 \times 10^6 \text{ nucleosomes.}$$
   (b) The average number of nucleosomes per loop of this condensed DNA would be:
$$(1.2 \times 10^6 \text{ nucleosomes})/(2000 \text{ loops}) = 6.0 \times 10^2 \text{ or } 600 \text{ nucleosomes/loop.}$$

2. Since the exonucleases do not act on this DNA, it may be circular. In double-stranded DNA, the thymine/adenine ratio is close to 1.00. The ratio 1.36 suggests that normal A-T pairing is not present. One explanation is that the DNA is not only circular, but is also single stranded.

3. Circular DNA can exist in relaxed circles or in supercoiled forms. (See text Figure 20.12.) Supercoiled circular DNA is more compact than the same DNA in relaxed circles. Vinograd's DNA contained some of both relaxed and supercoiled forms. The more compact, supercoiled DNA sedimented more quickly and the relaxed form of the circular DNA sedimented less quickly, giving rise to the two DNA bands observed.

4. The somewhat random action of DNase I suggests that no specific base sequence in the substrate DNA is required by the enzyme. Lysine and arginine residues of the enzyme are positively charged and participate in charge-charge interactions with negatively-charged phosphate groups of the sugar-phosphate backbone of B-DNA. This is thought to occur in the minor groove of B-DNA, but cannot occur in Z-DNA because its minor groove is too narrow for enzyme access. Also, the sugar-phosphate backbone of Z-DNA is oriented more toward the interior of the molecule than is the case in B-DNA.

5. Cytosine is converted to uracil by nitrous acid and adenine is converted to hypoxanthine, as shown in Figure 20.2. Uracil does not hydrogen bond properly with guanine, the normal base pair partner of cytosine. Likewise, hypoxanthine does not hydrogen bond properly with thymine, the normal partner of adenine. These changes would cause errors in replication and transcription.

CHAPTER 21

A. True–False

1. <u>False.</u> The semiconservative model of DNA replication is supported by the experiments of Meselson and Stahl and others.

2. <u>False.</u> Replication of the circular, duplex DNA occurs in a bidirectional manner from a single initiation site.

3. <u>False.</u> Not the deoxyribonucleoside monophosphates, but the triphosphates (dATP, dGTP, dCTP, dTTP) are required. Pyrophosphate, PP_i, is released from each deoxyribonucleoside triphosphate during incorporation of the nucleotide monophosphate moiety into the DNA strand.

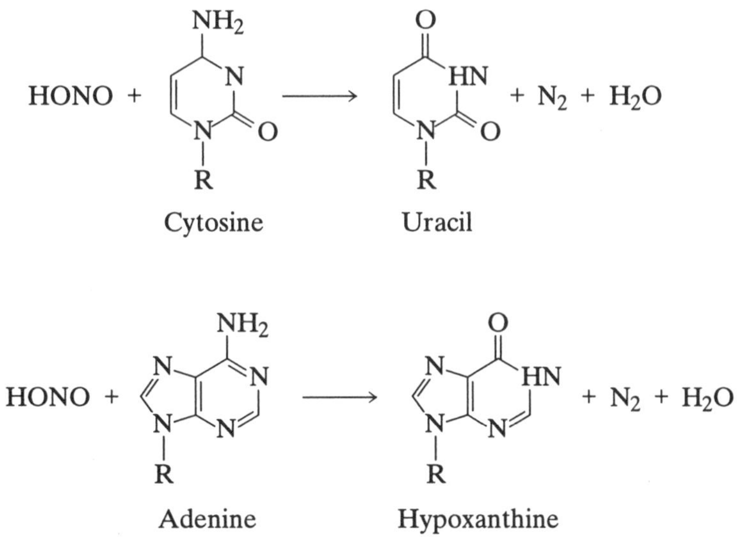

Figure 20.2 The Reaction of Nitrous Acid with Bases of DNA.

4. <u>True.</u> *E. coli* cells have three different DNA polymerases and eukaryotes have at least five different DNA polymerases.

5. <u>False.</u> DNA polymerase I also requires preformed DNA to serve both as a primer strand and as a template strand.

6. <u>True.</u> DNA polymerase III catalyzes chain elongation during DNA replication. It is a key component of the replisome.

7. <u>True.</u> Eukaryotic Okazaki fragments contain 100–200 nucleotide residues, but prokaryotic Okazaki fragments contain about 1000 nucleotide residues.

8. <u>True.</u> DNA ligase catalyzes the joining of one strand segment to another (seals nicks) by catalyzing the formation of phosphodiester bonds.

9. <u>False.</u> Thymine dimers are corrected by the action of DNA photolyase. The process is called photoreactivation and is a type of direct repair, not excision repair.

10. <u>True.</u> The RecA strand exchange protein recognizes regions of homology in the DNA single and duplex strands and promotes formation of a triple-stranded intermediate in the process of recombination.

B. Short Answers

1. replication fork
2. template
3. primers
4. leading strand; lagging strand
5. processive

6. Okazaki fragments
7. Nick translation
8. Holliday junction
9. replisome
10. helicases

C. Problems

1. Circular duplex DNA in the bacterial cell forms large loops and is highly supercoiled. It is, therefore, very compact.

2. The phosphate groups in DNA are ionized under cellular conditions, carry negative charges, and make the exterior of the helical molecule highly polar. Acid furnishes $H^{\oplus}$ ions that protonate the phosphate anions, rendering them nonionic and making the exterior much less polar and, therefore, less soluble in water.

3. The proofreading function of DNA polymerase III involves the $3' \rightarrow 5'$ exonuclease activity of the enzyme, which catalyzes the removal of a mismatched deoxyribonucleotide that has just been incorporated into a new DNA strand during replication.

4. As depicted in text Figure 21.14, there are two polymerase core complexes, one for each of the DNA template strands. The lagging strand template loops back through the replisome so that synthesis of both DNA strands occurs in the direction of movement of the replication fork.

5. In the final stages of lagging strand synthesis, RNA primers are removed by the $5' \rightarrow 3'$ exonuclease activity of DNA polymerase I, which then performs nick translation to fill in the gaps left by the RNA primers. Once these new DNA segments are in place, the sugar-phosphate backbone is sealed by the action of DNA ligase. Therefore, DNA samples isolated after replication contain no RNA.

6. $2',3'$-Dideoxyribonucleoside triphosphates are nucleoside triphosphates that lack oxygen at both the $2'$ and $3'$ positions on the ribose moiety. During DNA synthesis, they are unable to form a phosphodiester link at the $3'$ position as are normal deoxyribonucleoside triphosphates. Incorporation of a dideoxy form terminates chain synthesis. ddNTPs are used in the Sanger method of DNA sequencing, as discussed in text Section 21.6.

7. The damaged region is detected and an endonuclease cleaves the sugar-phosphate backbone on both sides of the damaged area, so that the oligonucleotide is then removed. The resulting gap is filled in by the action of DNA polymerase I or a similar enzyme, and the nick in the sugar phosphate backbone is sealed by the action of DNA ligase. (See text Section 21.8.B and text Figure 21.20.)

8. DNA is replicated in the nucleus of eukaryotic cells and proteins are synthesized in the cytosol. As indicated, the number of histones doubles with each round of replication. The newly formed histones are used in the formation of nucleosome and chromatin structures of daughter DNA molecules.

9. The *E. coli* chromosome contains 4.6×10^6 nucleotide pairs or 9.2×10^6 total bases, all of which must be replicated. One error occurs per 10^7 replication steps. The average number of incorrect bases is:

$$9.2 \times 10^6 \text{ polymerization steps} \times 1 \text{ error}/10^7 \text{ polymerization steps} = 0.92 \text{ incorrect bases.}$$

About one incorrect base is incorporated per replication of the chromosome.

10. (a) Both DNA ligases form an AMP-DNA ligase intermediate. The *E. coli* DNA ligase uses $NAD^\oplus$ as cosubstrate and supplier of the AMP moiety, while eukaryotic DNA ligase uses ATP for this function.

(b) The kinetic mechanism is ping-pong. The cosubstrate ($NAD^\oplus$ or ATP) binds first to the enzyme and the first product ($NMN^\oplus$ or PP_i) is then released before the next substrate (the nicked DNA) binds and reacts.

(c) The DNA ligases do not have a specificity for any particular base sequence in DNA. They recognize the gap between the $3'$-OH group of one ribose and the adjacent $5'$-phosphoryl group. (Note, in the caption of text Figure 21.13, "B" can be any of the bases found in DNA.)

D. Additional Problems

1. Gel electrophoresis of two such samples could help to identify the prokaryotic Okazaki fragments as being the longer, higher molecular weight fragments. The eukaryotic fragments would travel farther down the gel since they are typically only one- to two-tenths the size of prokaryotic Okazaki fragments.

2. (a) There are 4.6×10^6 nucleotide pairs in the *E. coli* chromosome but only one DNA strand (4.6×10^6 nucleotides) is replicated discontinuously to produce Okazaki fragments. Each Okazaki fragment (OF) is 1000 nucleotides in length. The number of fragments in one-third of the chromosome is:

$$4.6 \times 10^6 \text{ nucleotides} \times 1 \text{ OF}/10^3 \text{ nucleotides} \times 1/3 = 1530 \text{ or about 1500 OF.}$$

(b) The rate of new strand extension is 1000 nucleotides per second (text Section 21.1.). Nucleotides are incorporated simultaneously along the leading strand and the lagging strand at this rate. Replication is bidirectional, therefore, there are two replication forks moving in opposite directions on the circular DNA molecule making the overall rate 2×10^3 nucleotides per second. The time is:

$$4.6 \times 10^6 \text{ nucleotides/chromosome} \times 1 \text{ s}/(2 \times 10^3 \text{ nucleotides}) \times 1/3 \text{ chromosome} \times 1 \text{ min}/60 \text{ s} = 12.7 \text{ minutes.}$$

3. (a) The ^3H-labeled DNA is newly synthesized DNA that was formed by incorporation of the ^3H-deoxythymidine during replication. (The ^3H-deoxythymidine was first converted to ^3H-TTP by the cell.)

(b) Both short and long fragments of newly-synthesized DNA occurred because of the different modes of synthesis on the two existing DNA strands that serve as templates. Discontinuous synthesis on the lagging strand produced the short DNA fragments. (See text Figure 21.9.)

4. The direction of travel of the fragments in the gel shown in Figure 21.1 is from top to bottom. The smallest fragments travel the farthest down the gel. The smallest fragment will be that formed by the attachment of one base (the ddNTP) to the primer. The third lane contains the smallest fragment, which must be 5´-TAC-3´, since ddCTP was used in that trial. Using this reasoning, the sequence of bases that attach to the primer can be read from bottom to top of the gel. That sequence, including the primer, is 5´-*TA*CCAATGGC-3´. The complementary DNA strand, which served as a template for synthesis of the strand attached to the primer, must be 5´-GCCATTGGTA-3´. The duplex DNA segment is shown here with the primer in italics.

5´-*TA*CCAATGGC-3´ (synthesized strand)

3´-ATGGTTACCG-5´ (existing strand)

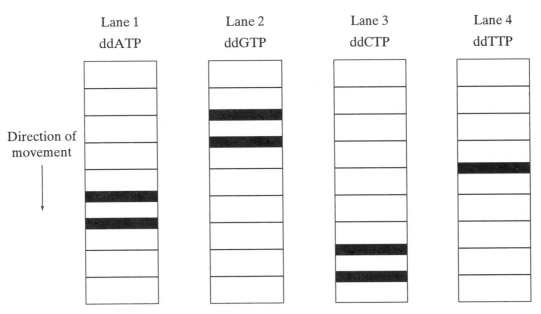

Figure 21.1 Sanger's DNA Sequencing Method.

5. The fragments are listed from smallest to the largest in each lane.

Lane 1:   5´-TACCA-3´; 5´-TACCAA-3´

Lane 2:   5´-TACCAATG-3´; 5´-TACCAATGG-3´

Lane 3:   5´-TAC-3´; 5´-TACC-3´; 5´-TACCAATGGC-3´

Lane 4:   5´-TACCAAT-3´

CHAPTER 22

A. True–False

1. <u>False.</u> Modification of the four common bases occurs after polymerization (i. e., transcription) to produce these unusual bases.

2. <u>False.</u> Promoters are double-stranded DNA sequences. RNA polymerase binds to both strands during transcription initiation.

3. <u>True.</u> The housekeeping genes are said to be constitutively expressed. They are not inducible.

4. <u>True.</u> Translation is coupled to transcription in prokaryotes.

5. <u>True.</u> The poly A tails on the 3´ end help prevent exonuclease cleavage of the coding region of the mature mRNA.

6. <u>False.</u> This cap is added to the 5´ end of the mRNA shortly after it is synthesized. The poly A tail is added to the 3´ end of the mRNA.

7. <u>True.</u> During processing, spacer sequences are removed from the large precursor rRNA molecule to yield the individual rRNA molecules that are components of eukaryotic ribosomes.

8. <u>True.</u> The σ^{70} subunit is mainly responsible for formation of an initiation complex at the site of σ^{70}-specific promoters.

9. <u>False.</u> Three RNA polymerase complexes catalyze transcription in the cell nucleus, but different RNA polymerase complexes are also found in mitochondria and in chloroplasts of those eukaryotes that contain mitochondria and chloroplasts. (See text Table 22.4 for more details.)

10. <u>True.</u> The process of splicing is catalyzed by a large RNA complex called the spliceosome.

B. Short Answers

1. promoter

2. RNA polymerase holoenzyme (Note that the holoenzyme includes the sigma factor which dissociates from the holoenzyme after transcription is initiated. The core polymerase, consisting of the $\alpha_2\beta\beta'$ subunits, catalyzes elongation of the RNA molecule.)

3. transcription bubble

4. rho Factor

5. RNA polymerase II

6. introns; exons (These terms also refer to the regions of the gene (DNA) than encode corresponding RNA introns and exons.)

7. small nuclear RNA's (snRNA's)

8. repressor

9. activator

10. inducers

C. Problems

1. The TATA box is a region of consensus sequence, TATAAT, that is part of some promoters. It is part of the DNA sequence to which RNA polymerase binds during the initiation of transcription.

2. The protein rho binds to single-stranded RNA behind the paused transcription complex, promotes disruption of the transcription complex and its dissociation from the DNA template strand.

3. The *lacI* gene is located upstream from the *lac* operon. It encodes a regulatory protein called *lac* repressor that binds to DNA near the promoter of the *lac* operon and prevents transcription.

4. In catabolite repression, one catabolite represses the expression of enzymes required for the catabolism of other catabolites. For example, in a growth medium that contains glucose along with other catabolites (for example, lactose), *E. coli* preferentially uses glucose. In this case, glucose causes catabolite repression. In the absence of glucose, levels of cAMP rise resulting in the activation of genes that are normally not transcribed at high rates in the presence of glucose.

5. CRP (cAMP regulatory protein) is a prokaryotic transcriptional regulatory protein. It binds cAMP to form a complex that can act as an activator of genes subject to catabolite repression.

6. RNA polymerase I catalyzes the synthesis of large precursor rRNA molecules.

   RNA polymerase II catalyzes the synthesis of precursor mRNA molecules and transcribes some genes that encode small RNA molecules.

   RNA polymerase III catalyzes the synthesis of tRNA, 5S rRNA, and some other small RNA molecules.

7. Capping of eukaryotic mRNA molecules at their 5′ ends with 7-methylguanylate groups protects the mRNA molecules from degradation by exonucleases. It also promotes proper binding of mRNA molecules to ribosomes in preparation for protein synthesis.

8. Text Section 22.2.B indicates that *E. coli* transcription rates range from 30 to 85 nucleotides per second. Selecting an average of 58 nucleotides per second, the time required would be $(4.6 \times 10^6 \text{ bp})(0.978) \times 1 \text{ s}/58 \text{ bp} \times 1 \text{ hr}/3600 \text{ s} = 21.5$ hr or about one day for a single RNA polymerase molecule to transcribe 97.8% of the *E. coli* chromosome.

9. (a) The error rate in RNA synthesis is higher than that of DNA replication because RNA polymerase does not have the exonuclease proofreading activity possessed by DNA polymerase III.

(b) In some cases (sickle cell anemia), a single amino acid substitution in a protein can be disadvantageous. If the error is in the DNA, it affects all molecules of the encoded protein. If an error occurs in synthesizing an mRNA molecule, the problem resides in only that one molecule and not in other copies of the mRNA, so only a small fraction of protein molecules are defective.

10. In regions of dyad symmetry, a segment of one DNA strand contains the same sequence of bases as a nearby segment in the other DNA strand. These segments are not hydrogen bonded to each other, but are separated in the duplex DNA by a short segment (eg., four base pairs, as shown in text Figure 22.8). Such a region aids or exaggerates a pause in transcription because the transcript RNA molecule forms a hairpin that may destabilize the DNA-RNA transcription hybrid.

D. Additional Problems

1. Figure 22.1 illustrates the induction of the *lac* operon by IPTG since *lac* mRNA begins to appear after IPTG is added to the medium. The operon is transcribed at a high rate in the absence of glucose because CRP-cAMP is present. The addition of glucose lowers cAMP levels thus abolishing activation by CRP-cAMP. This illustrates catabolite repression.

2. The original concentration of R is (10 molecules/cell) $\times$ (1 mol/6.02×10^{23} molecules) $\times$ (1 cell/0.30×10^{-15} L) = 5.5×10^{-8} M.

   The original concentration of O is (2 molecules/cell) $\times$ (1 mol/6.02×10^{23} molecules) $\times$ 1 cell/0.30×10^{-15} L) = 1.1×10^{-8} M.

   At equilibrium, K_D = [R][O]/[RO] = 10^{-13} M. Practically all O is complexed due to the high association constant (K_{Assoc} = 10^{13} M^{-1}). Thus, we can estimate the concentration of RO as 1.1×10^{-8} M. The equilibrium concentration of R will be 5.5×10^{-8} M -1.1×10^{-8} M = 4.4×10^{-8} M. Then,

   [O] = K_D[RO]/[R] = (10^{-13})(1.1×10^{-8})/(4.4×10^{-8}) = 2.5×10^{-14} M.

   Compared to the original concentration of O, 99.9998% of O is bound by R at equilibrium.

3. The data in Figure 22.2 show that three components of the mixture were detected by their radioactivity. Most of the RNA was found in the first fractions, which were more dense and sedimented toward the bottom of the CsCl gradient. The second RNA fraction occurs at about the same location as does the majority of the less dense, denatured DNA. This was taken as evidence of a DNA-RNA hybrid that should form if their sequences were complementary, as they would be if one DNA strand had served as a template for synthesis of the RNA. This fraction containing the hybrid does not occur between that of the nonhybridized RNA and DNA because the RNA segment in the hybrid is short compared to the length of the DNA strand with which it hydrogen bonds. Although not shown in Figure 22.2, no DNA-RNA hybrid was formed unless the DNA was first denatured.

4. After about 10 nucleotides are incorporated into the RNA chain, the holoenzyme undergoes a conformational change that causes release of the σ^{70} subunit. The core polymerase no longer binds the promoter sequence any more strongly than other DNA sequences and is able to move away from the promoter (promoter clearance). Accessory proteins, such as Nus A, may also be involved in this process.

5. Not removing an intron region from an mRNA precursor would produce an aberrant mRNA with a disrupted coding frame. The resulting protein might contain additional amino acids or translation might terminate prematurely due to the presence of in-frame stop codons in the intron.

CHAPTER 23

A. True–False

1. <u>False.</u>   The aminoacyl-tRNA synthetases bind specific tRNA's as well.

2. <u>True.</u>   This is also true in eukaryotic systems.

3. <u>False.</u>   The amino acids are covalently attached to the 3′ end of the tRNA (to the 2′- or 3′-hydroxyl group of the adenylate residue of the terminal CCA sequence).

4. <u>False.</u>   It does so in many prokaryotes, but in eukaryotes, methionine is not formylated.

5. <u>True.</u>   The Shine-Delgarno sequence, located upstream of the initiator codon, is the ribosomal binding site.

6. <u>False.</u> The peptidyl transferase activity is located in the 50S ribosomal subunit and appears to be localized in the 23S rRNA molecule of the ribosomal subunit. (RNA molecles that catalyze reactions are called ribozymes by some and RNA enzymes by others.)

7. <u>True.</u> Since the peptide chain grows from the amino toward the carboxy terminus, the amino group of the new amino acid being incorporated must be free to react.

8. <u>False.</u> Prokaryotic mRNA molecules may be polycistronic, but those of eukaryotes generally are not.

9. <u>True.</u> Eukaryotic mRNA molecules do not have the Shine-Delgarno sequences, but they do have the 5´-cap that helps in the binding of the small ribosomal unit.

10. <u>False.</u> Such proteins are transported into the lumen of the endoplasmic reticulum, then to the Golgi apparatus, and then into vesicles for transport to the plasma membrane for secretion. Posttranslational modification of these proteins can occur in each of these three locations.

B. Short Answers

1. Isoacceptor tRNA molecules

2. phenylalanine; UUU

3. proofreading

4. acceptor stem

5. synonymous codons

6. translocation

7. polysome (or polyribosome)

8. four

9. signal peptide

10. release factors

C. Problems

1. An amino acid is made ready for incorporation into a protein by its covalent attachment to a tRNA molecule specific for that amino acid. An aminoacyl adenylate intermediate reacts with the 3´-OH group of the terminal adenylate group of the tRNA to form an ester linkage. The reaction is catalyzed by an aminoacyl-tRNA synthetase specific for both the amino acid and the tRNA molecule. The reactions are outlined in text Figure 23.9.

2. No bacterial mRNA was being produced in this system, since there were no intact cells. mRNA is short-lived, so there probably was very little in the supernatant fraction. This experiment showed that mRNA provided the templates necessary for protein synthesis.

3. Assuming a triplet code, one could conclude:
   a. a code for lysine is AAA.
   b. a code for asparagine is AAC.
   c. mRNA templates are translated in a 5´→3´ direction.
   d. proteins are synthesized from their amino terminus toward their carboxy terminus.

4. Yes, this allows the bacterial cell to start synthesis of proteins encoded by a particular mRNA molecule even before synthesis of the mRNA molecule is complete.

5. The term degenerate refers to the fact that there is more than one codon for most amino acids. Synonymous codons are those that encode the same amino acid.

6. (a) If the amino acid was recognized by the codon, the radioactive alanine would have been incorporated into some of the hemoglobin sites normally occupied by alanine.

   (b) If the tRNA was recognized by the codon, radioactive alanine would have been incorporated into hemoglobin sites normally occupied by cysteine. (The latter result was obtained by Chapeville and coworkers, who originally performed this study.)

7. (a) The formyl group of *N*-formylmethionine protects the amino group of methionine and prevents the derivatized amino acid from being incorporated anywhere other than at the N-terminus of the chain. In addition, *N*-formylmethionine is only found on initiator tRNA's, which are not recognized by EF-Tu. Thus, the formylated methionine is never bound in the ribosomal A-site.

(b) *N*-formylmethionyl-tRNA$_f^{Met}$ is the only charged tRNA that binds with the 30S ribosomal subunit to form the 30S initiation complex, so methionyl-tRNA^{Met} cannot initiate protein synthesis.

8. (a) Elongation factor EF-G forms a complex with GTP that binds to the 70S ribosomal complex. Hydrolysis of GTP provides the energy needed to release the uncharged tRNA from the P-site and to cause movement of the mRNA with respect to the ribosome so that the next codon is located at the A-site. The EF-G-GDP complex then dissociates from the ribosomal complex and the next aminoacyl-tRNA can enter the A site.

   (b) The process is called translocation.

9. (a) While the basic process is the same in both systems, the initiator methionyl-tRNA$_i^{Met}$ does not require formylation in eukaryotes as is the case in many prokaryotes.

   (b) Eukaryotic mRNA molecules encode a single protein while those of many prokaryotes encode several proteins.

   (c) Eukaryotic mRNA molecules have no Shine-Delgarno sequence for binding ribosomes, but do have 7-methylguanylate caps at the 5′-end that aid in binding the 40S ribosomal subunit.

   (d) More protein factors are involved in the eukaryotic initiation and elongation processes than are involved in the corresponding prokaryotic processes.

10. The incorporation of one amino acid residue into a protein chain consumes the equivalent of four phosphoanhydride bonds. Use the $\Delta G^{0'}$ value of 30. kJ/mol for phosphoanhydride bond synthesis to calculate the energy involved in the synthesis of 0.685 mole of a protein that contains 250 amino acid residues:

$$0.685 \text{ mol protein} \times 250 \text{ residues/mol protein} \times 4 \text{ phosphoanhydride bonds/residue}$$
$$\times 30. \text{ kJ/(mol phosphoanhydride bond)} = 2.1 \times 10^4 \text{ kJ.}$$

D. Additional Problems

1. (a) Synthesis of the 300-residue polypeptide requires at least 900 nucleotides (NT) in the mRNA. If produced in 20 seconds, the nucleotides read per second are:

$$(300 \text{ residues} \times 3 \text{ NT/residue})/20 \text{ sec} = 45 \text{ NT/sec}$$

   (b) If one protein was produced each 20 seconds by a single ribosome for each of 5000 mRNA molecules, the number of proteins produced per second would be:

$$(5000 \text{ mRNA} \times 1 \text{ protein/mRNA/ribosome})/20 \text{ sec} = 250 \text{ proteins/ribosome/sec.}$$

   If, however, the cell is producing 1000 proteins per second, the number of ribosomes reading a particular mRNA is:

$$(1000 \text{ proteins/sec})/(250 \text{ proteins/ribosome/sec}) = 4 \text{ ribosomes.}$$

2. (a) The translation process inhibited was initiation. The Shine-Delgarno sequences in prokaryotic mRNA molecules aid in binding the 30S ribosomal subunit. Binding of the oligonucleotides by 30S ribosomal subunits prevented them from initiating protein synthesis.

   (b) Eukaryotic protein synthesis was not affected because initiation of translation of eukaryotic mRNA molecules does not involve binding to Shine-Delgarno sequences.

   (c) No proteins were formed by those ribosomes bound by the oligonucleotides. Ribosomes not bound by the oligonucleotides produced normal protein chains. The overall rate of protein synthesis was decreased.

3. (a) The following table gives the codons possible using only uracil and guanine bases, the amino acids for which they code, and the expected probability for each, expressed as a percentage. Percentages were determined as follows: For UUU, $(0.76 \times 0.76 \times 0.76) (100) = 44\%$.

Codon	Amino Acid	Probability
UUU	Phe	44%
UUG	Leu	14%
UGU	Cys	14%
GUU	Val	14%
GGU	Gly	4%
GUG	Val	4%
UGG	Trp	4%
GGG	Gly	1%

(b) In a polypeptide of 1000 residues, there would likely be 440 phenylalanine residues, on average.

(c) In a polypeptide of 1000 residues, there would likely be 40 tryptophan residues, on average.

4. (a) Proteins destined for secretion contain an N-terminal segment called the signal peptide. The signal peptide is recognized and bound by the protein-RNA complex called a signal-recognition particle. Hence, the name.

   (b) The SRP-ribosome binds to a receptor protein on the endoplasmic reticulum and the signal peptide is inserted into the membrane of the endoplasmic reticulum. This binding of the ribosome to the endoplasmic reticulum relieves the inhibition and synthesis of the protein can continue.

5. (a) In the modified residue, oxygen is replaced by selenium.

   (b) The UGA codon is normally a stop codon.

CHAPTER 24

A. True–False

1. <u>False.</u> They are circular, *double-stranded*, supercoiled DNA molecules.

2. <u>False.</u> Recombinant DNA molecules are made in nature, for instance, during infection of a host cell by a bacteriophage or virus.

3. <u>True.</u> The vector acts as the carrier molecule for the inserted DNA fragment.

4. <u>True.</u> This is possible by the use of the enzyme reverse transcriptase. The DNA so made is referred to as cDNA (complementary DNA).

5. <u>True.</u> Such vectors are called shuttle vectors.

6. <u>False.</u> The complementary sticky ends will hydrogen bond with each other, but the sugar phosphate backbones of the two DNA segments must be sealed by the action of DNA ligase.

7. <u>False.</u> The direct uptake of recombinant DNA molecules by a host cell is called genetic transformation. Recombinant DNA can also be packaged into virus particles and transferred to the host cells by the process of transfection.

8. <u>False.</u> The transformants would be sensitive to ampicillin, but not to tetracycline. See text Section 24.3.A.

9. <u>False.</u> The stuffer fragment is not used. The λ arms are purified and combined with the genomic DNA to create the recombinant molecules to be packaged into phage particles.

10. <u>True.</u> The poly A tails are recognized by synthetic oligo dT molecules that associate with the poly A tail regions, by base pairing, and serve as primers for the synthesis of complementary DNA strands, catalyzed by reverse transcriptase.

B. Short Answers

1. transformation
2. library
3. Gene therapy
4. probe
5. expression vectors

6. transgenic
7. restriction fragment length polymorphisms (RFLPs)
8. polymerase chain reaction
9. Insertional inactivation
10. transfection

C. Problems

1. Four types of vectors that have been used are plasmids, bacteriophages, viruses, and artificial chromosomes, such as YAC's. (One might also include cosmids, which are special types of plasmids that carry the cos sites (nucleotide sequences of the cohesive ends) of phage lambda DNA.)

2. Subjecting the solution containing the cDNA-mRNA hybrid to base-catalyzed hydrolysis would cause degradation of the mRNA chain, but would leave the cDNA strand intact.

3. In cases where the insert DNA is large, plasmids are not the best choice, since they transform poorly. λ Phage is the preferred vector in such cases.

4. (a) DNA cannot be packaged into phage particles to be used for transfer to host cells if it is not of the appropriate size. To be packaged, such DNA must be about 45 to 50 kb long.

   (b) λ Phage DNA must be engineered so that a sizeable piece can be removed and replaced by foreign DNA. The phage DNA is cleaved into stuffer DNA fragments and λ arms. The λ arms are used to make the recombinant DNA, but the stuffer fragments are not used.

5. DNA polymerase I has the $5' \rightarrow 3'$ exonuclease activity necessary to remove the mRNA segments that remain after use of RNase H. It also has the activity required for the synthesis of DNA to replace these RNA segments. DNA polymerase III does not have the required exonuclease activity.

6. Since cDNA libraries are constructed from mature mRNA molecules, the DNA contains only expressed genes. It does not contain introns, flanking sequences, or regulatory sequences that might be of interest to some researchers. The cDNA libraries may, however, be more difficult to construct if they are designed to represent all mRNA molecules, including those of both high and low abundance.

7. Site-directed mutagenesis of the gene that codes for the enzyme allows the alteration of a single nucleotide sequence in the gene. This leads to the replacement of a single amino acid in an enzyme by another amino acid. If the replaced amino acid is involved in the catalytic activity of the enzyme, its replacement by another amino acid residue will very likely have an effect on the catalytic activity.

8. (a) The problem of elevated temperatures is overcome by the use of a heat-stable DNA polymerase for synthesis of the new DNA strands. The DNA polymerase from *Thermus aquaticus, Taq* polymerase, is stable to temperatures above 90°C.

   (b) Amplification of small amounts of DNA provides enough for cloning, use as a screening probe, or sequencing. The technique has found use in the study of DNA from preserved samples (120 million-year old insects), in forensic analysis, and screening of individuals for genetic abnormalities.

9. Researchers have done this by synthesizing a short duplex DNA molecule that contains a desired restriction site. This synthetic DNA is then attached to the DNA molecule of interest (with blunt ends) by the action of T4 DNA ligase. The resulting molecule has a tailor-made restriction site that will provide sticky ends when cleaved.

10. By use of a hybrid promoter that requires a specific exogenous activator, expression of the inserted gene can be easily controlled. Bacterial cells could be cloned with the hybrid promoter turned off so that energy is not wasted by gene expression. Once sufficient bacterial colonies exist, expression of the protein product could be induced by addition of the activator. In addition, the eukaryotic protein might kill the bacterial cell, so that it should be made only when cells are ready to be harvested.

D. Additional Problems

1. Expression of the recombinant DNA molecule gave a protein product that included somatostatin attached by a methionine residue to the protein product from the *lac* gene. (AUG is the codon for methionine.) This allowed the protein product to be cleaved by BrCN to yield somatostatin itself.

2. It is easy to deduce the sequence of a DNA strand that served as a template for a particular mRNA by writing the sequence of bases that would form complementary base pairs with those of the mRNA. The sequence of the corresponding DNA coding strand is the same as that of the mRNA except for the use of thymine in locations where uracil appears in the mRNA. However, because of the degeneracy of the genetic code, several possible mRNA (and, therefore, several DNA) sequences could give rise to a particular amino acid sequence.

3. Since Trp and Met have only one codon each, but Phe, Lys, and Glu each have two codons, there are eight possible probe sequences.

4. The codons for the amino acids of the peptide are:

Trp	Phe	Lys	Glu	Met
UGG	UUU	AAA	GAA	AUG
	UUC	AAG	GAG	

Sequences for the DNA coding strand probes will be the same as those of the mRNA strands with thymine bases used instead of uracil bases. (Note: Dashes are used only for clarity.)

   1. TGG-TTT-AAA-GAA-ATG
   2. TGG-TTC-AAA-GAA-ATG
   3. TGG-TTT-AAG-GAA-ATG
   4. TGG-TTT-AAG-GAG-ATG
   5. TGG-TTT-AAA-GAG-ATG
   6. TGG-TTC-AAA-GAG-ATG
   7. TGG-TTC-AAG-GAA-ATG
   8. TGG-TTC-AAG-GAG-ATG

5. If $(10^3 \text{ bp})(100)/n \text{ bp} = 0.000031$, then $n = 10^5 \text{ bp}/(3.1 \times 10^{-5}) = 3.1 \times 10^9 \text{ bp} = 3.1$ billion base pairs.

6. Use text equation 24.1, with $P = 0.99$.

$$N = \ln(1-P)/\ln(1-n)$$

   Here, n is the ratio of the size of the insert to the size of the genome, or $10 \text{ kb}/144{,}000 \text{ kb} = 6.9 \times 10^{-5}$. Therefore,

$$N = \ln(1 - 0.99)/\ln(1 - 6.9 \times 10^{-5}) = \ln(0.01)/\ln(0.999931) = 66{,}739.$$

   Therefore, about 6.7×10^4 clones would need to be screened.

Appendix

Table A·1 Some physical constants used in biochemistry

R	Universal gas constant	$8.315 \text{ J K}^{-1} \text{ mol}^{-1}$
$\mathcal{F}$	Faraday's constant	$96.48 \text{ kJ V}^{-1} \text{ mol}^{-1}$
N	Avogadro's number	$6.022 \times 10^{23} \text{ mol}^{-1}$

Table A·2 Greek alphabet

A	α	alpha	N	ν	nu
B	β	beta	Ξ	ξ	xi
Γ	γ	gamma	O	o	omicron
Δ	δ	delta	Π	π	pi
E	ε	epsilon	P	ρ	rho
Z	ζ	zeta	Σ	σ	sigma
H	η	eta	T	τ	tau
Θ	θ	theta	Y	υ	upsilon
I	ι	iota	Φ	ϕ	phi
K	κ	kappa	X	χ	chi
Λ	λ	lambda	Ψ	ψ	psi
M	μ	mu	Ω	ω	omega

Table A·3 Units commonly used in biochemistry

Physical quantity	SI unit	Symbol
Length	Meter	m
Mass	Kilogram	kg
Energy	Joule	J
Electric potential	Volt	V
Time	Second	s
Temperature	Kelvin*	K

*273 K = 0°C.

Table A·4 Prefixes commonly used with SI units

Prefix	Symbol	Multiplication factor
Mega	M	10^{6}
Kilo	k	10^{3}
Milli	m	10^{-3}
Micro	μ	10^{-6}
Nano	n	10^{-9}
Pico	p	10^{-12}
Femto	f	10^{-15}

Table A·5 Atomic numbers and weights of the elements

Element	Symbol	Atomic number	Atomic weight	Element	Symbol	Atomic number	Atomic weight
Actinium	Ac	89	227.03	Mendelevium	Md	101	255.09
Aluminum	Al	13	26.98	Mercury	Hg	80	200.59
Americium	Am	95	243.06	Molybdenum	Mo	42	95.94
Antimony	Sb	51	121.75	Neodymium	Nd	60	144.24
Argon	Ar	18	39.95	Neon	Ne	10	20.18
Arsenic	As	33	74.92	Neptunium	Np	93	237.05
Astatine	At	85	210.99	Nickel	Ni	28	58.71
Barium	Ba	56	137.34	Niobium	Nb	41	92.91
Berkelium	Bk	97	247.07	Nitrogen	N	7	14.01
Beryllium	Be	4	9.01	Nobelium	No	102	255
Bismuth	Bi	83	208.98	Osmium	Os	76	190.20
Boron	B	5	10.81	Oxygen	O	8	16.00
Bromine	Br	35	79.90	Palladium	Pd	46	106.40
Cadmium	Cd	48	112.40	Phosphorus	P	15	30.97
Calcium	Ca	20	40.08	Platinum	Pt	78	195.09
Californium	Cf	98	249.07	Plutonium	Pu	94	242.06
Carbon	C	6	12.01	Polonium	Po	84	208.98
Cerium	Ce	58	140.12	Potassium	K	19	39.10
Cesium	Cs	55	132.91	Praseodymium	Pr	59	140.91
Chlorine	Cl	17	35.45	Promethium	Pm	61	145
Chromium	Cr	24	52.00	Protactinium	Pa	91	231.04
Cobalt	Co	27	58.93	Radium	Ra	88	226.03
Copper	Cu	29	63.55	Radon	Rn	86	222.02
Curium	Cm	96	245.07	Rhenium	Re	75	186.20
Dysprosium	Dy	66	162.50	Rhodium	Rh	45	102.91
Einsteinium	Es	99	254.09	Rubidium	Rb	37	85.47
Erbium	Er	68	167.26	Ruthenium	Ru	44	101.07
Europium	Eu	63	151.96	Samarium	Sm	62	150.40
Fermium	Fm	100	252.08	Scandium	Sc	21	44.96
Fluorine	F	9	18.99	Selenium	Se	34	78.96
Francium	Fr	87	223.02	Silicon	Si	14	28.09
Gadolinium	Gd	64	157.25	Silver	Ag	47	107.87
Gallium	Ga	31	69.72	Sodium	Na	11	22.99
Germanium	Ge	32	72.59	Strontium	Sr	38	87.62
Gold	Au	79	196.97	Sulfur	S	16	32.06
Hafnium	Hf	72	178.49	Tantalum	Ta	73	180.95
Helium	He	2	4.00	Technetium	Tc	43	98.91
Holmium	Ho	67	164.93	Tellurium	Te	52	127.60
Hydrogen	H	1	1.01	Terbium	Tb	65	158.93
Indium	In	49	114.82	Thallium	Tl	81	204.37
Iodine	I	53	126.90	Thorium	Th	90	232.04
Iridium	Ir	77	192.22	Thulium	Tm	69	168.93
Iron	Fe	26	55.85	Tin	Sn	50	118.69
Khurchatovium	Kh	104	260	Titanium	Ti	22	47.90
Krypton	Kr	36	83.80	Tungsten	W	74	183.85
Lanthanum	La	57	138.91	Uranium	U	92	238.03
Lawrencium	Lr	103	256	Vanadium	V	23	50.94
Lead	Pb	82	207.20	Xenon	Xe	54	131.30
Lithium	Li	3	6.94	Ytterbium	Yb	70	173.04
Lutetium	Lu	71	174.97	Yttrium	Y	39	88.91
Magnesium	Mg	12	24.31	Zinc	Zn	30	65.37
Manganese	Mn	25	54.94	Zirconium	Zr	40	91.22

Table A·6 General formulas of organic compounds, important functional groups, and linkages common in biochemistry

(a) *Organic compounds*

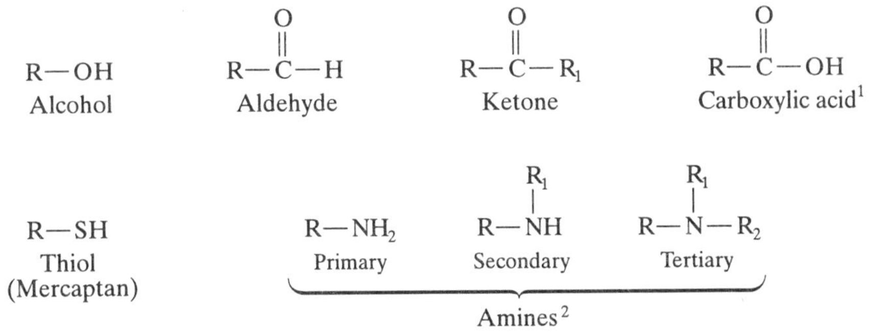

R—OH
Alcohol

$$R-\overset{\overset{\displaystyle O}{\|}}{C}-H$$
Aldehyde

$$R-\overset{\overset{\displaystyle O}{\|}}{C}-R_1$$
Ketone

$$R-\overset{\overset{\displaystyle O}{\|}}{C}-OH$$
Carboxylic acid[1]

R—SH
Thiol
(Mercaptan)

R—NH$_2$
Primary

$$R-\overset{\overset{\displaystyle R_1}{|}}{N}H$$
Secondary

$$R-\overset{\overset{\displaystyle R_1}{|}}{N}-R_2$$
Tertiary

Amines[2]

(b) *Functional groups*

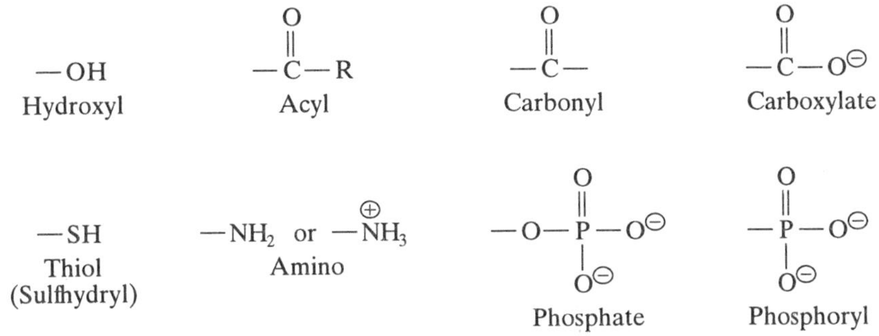

—OH
Hydroxyl

$$-\overset{\overset{\displaystyle O}{\|}}{C}-R$$
Acyl

$$-\overset{\overset{\displaystyle O}{\|}}{C}-$$
Carbonyl

$$-\overset{\overset{\displaystyle O}{\|}}{C}-O^{\ominus}$$
Carboxylate

—SH
Thiol
(Sulfhydryl)

—NH$_2$ or —$\overset{\oplus}{N}$H$_3$
Amino

$$-O-\overset{\overset{\displaystyle O}{\|}}{\underset{\underset{\displaystyle O^{\ominus}}{|}}{P}}-O^{\ominus}$$
Phosphate

$$-\overset{\overset{\displaystyle O}{\|}}{\underset{\underset{\displaystyle O^{\ominus}}{|}}{P}}-O^{\ominus}$$
Phosphoryl

(c) *Linkages*

$$-\overset{|}{\underset{|}{C}}-O-\overset{\overset{\displaystyle O}{\|}}{C}-$$
Ester

$$-\overset{|}{\underset{|}{C}}-O-\overset{|}{\underset{|}{C}}-$$
Ether

$$-\overset{|}{N}-\overset{\overset{\displaystyle O}{\|}}{C}-$$
Amide

$$-\overset{|}{\underset{|}{C}}-O-\overset{\overset{\displaystyle O}{\|}}{\underset{\underset{\displaystyle O^{\ominus}}{|}}{P}}-O^{\ominus}$$
Phosphate ester

$$-O-\overset{\overset{\displaystyle O}{\|}}{\underset{\underset{\displaystyle O^{\ominus}}{|}}{P}}-O-\overset{\overset{\displaystyle O}{\|}}{\underset{\underset{\displaystyle O^{\ominus}}{|}}{P}}-O-$$
Phosphoanhydride

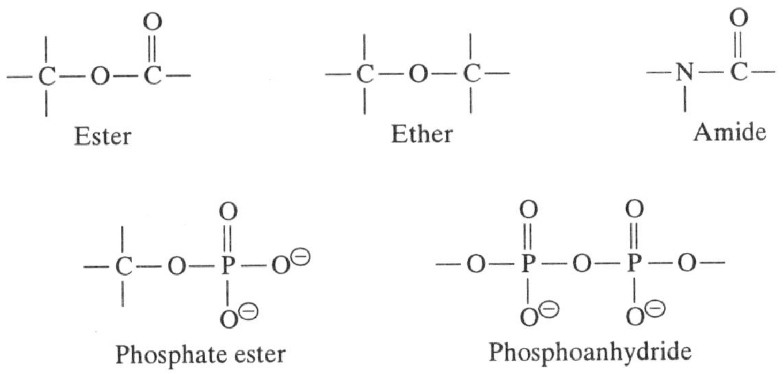

[1] Under most biological conditions, carboxylic acids exist as carboxylate anions: $R-\overset{\overset{\displaystyle O}{\|}}{C}-O^{\ominus}$.

[2] Amines can also be protonated: $R-\overset{\oplus}{N}H_3$, $R-\overset{\overset{\displaystyle R_1}{|}}{\overset{\oplus}{N}}H_2$, and $R-\overset{\overset{\displaystyle R_1}{|}}{\overset{\oplus}{N}}H-R_2$.

Table A·7 Dissociation constants and pK_a values of weak acids in aqueous solutions at 25°C

Acid	K_a (M)	pK_a
HCOOH (Formic acid)	1.77×10^{-4}	3.75
CH_3COOH (Acetic acid)	1.76×10^{-5}	4.75
$CH_3CHOHCOOH$ (Lactic acid)	1.37×10^{-4}	3.86
H_3PO_4 (Phosphoric acid)	7.52×10^{-3}	2.12
$H_2PO_4^{\ominus}$ (Dihydrogen phosphate ion)	6.23×10^{-8}	7.21
$HPO_4^{2\ominus}$ (Monohydrogen phosphate ion)	2.20×10^{-13}	12.70
H_2CO_3 (Carbonic acid)	4.30×10^{-7}	6.37
$HCO_3^{\ominus}$ (Bicarbonate ion)	5.61×10^{-11}	10.30
$NH_4^{\oplus}$ (Ammonium ion)	5.62×10^{-10}	9.25
$CH_3NH_3^{\oplus}$ (Methylammonium ion)	2.70×10^{-11}	10.60

Table A·8 pK_a values of some commonly used buffers

Buffer	pK_a at 25°C
Phosphate (pK_1)	2.2
Acetate	4.8
MES (2-(N-Morpholino)ethanesulfonic acid)	6.1
Citrate (pK_3)	6.4
PIPES (Piperazine-N,N'-bis (2-ethanesulfonic acid))	6.8
Phosphate (pK_2)	7.2
HEPES (N-2-Hydroxyethylpiperazine-N'-2-ethanesulfonic acid)	7.5
Tris (Tris (hydroxymethyl) aminomethane)	8.1
Glycylglycine	8.2
Glycine (pK_2)	9.8

[Adapted from Stoll, V. S., and Blanchard, J. S. (1990). Buffers: principles and practice. *Methods Enzymol.* 182:24–38.]

Table A·9 Some common fatty acids (anionic forms) incorporated in membrane lipids

Number of carbons	Number of double bonds	Common name	IUPAC name	Melting point, °C	Molecular formula
12	0	Laurate	Dodecanoate	44	$CH_3(CH_2)_{10}COO^{\ominus}$
14	0	Myristate	Tetradecanoate	52	$CH_3(CH_2)_{12}COO^{\ominus}$
16	0	Palmitate	Hexadecanoate	63	$CH_3(CH_2)_{14}COO^{\ominus}$
18	0	Stearate	Octadecanoate	70	$CH_3(CH_2)_{16}COO^{\ominus}$
20	0	Arachidate	Eicosanoate	75	$CH_3(CH_2)_{18}COO^{\ominus}$
22	0	Behenate	Docosanoate	81	$CH_3(CH_2)_{20}COO^{\ominus}$
24	0	Lignocerate	Tetracosanoate	84	$CH_3(CH_2)_{22}COO^{\ominus}$
16	1	Palmitoleate	cis-Δ^9-Hexadecenoate	− 0.5	$CH_3(CH_2)_5CH{=}CH(CH_2)_7COO^{\ominus}$
18	1	Oleate	cis-Δ^9-Octadecenoate	13	$CH_3(CH_2)_7CH{=}CH(CH_2)_7COO^{\ominus}$
18	2	Linoleate	cis, cis-$\Delta^{9,12}$-Octadecadienoate	− 9	$CH_3(CH_2)_4(CH{=}CHCH_2)_2(CH_2)_6COO^{\ominus}$
18	3	Linolenate	all cis-$\Delta^{9,12,15}$-Octadecatrienoate	−17	$CH_3CH_2(CH{=}CHCH_2)_3(CH_2)_6COO^{\ominus}$
20	4	Arachidonate	all cis-$\Delta^{5,8,11,14}$-Eicosatetraenoate	−49	$CH_3(CH_2)_4(CH{=}CHCH_2)_4(CH_2)_2COO^{\ominus}$

[Values from Dawson, R. M. C., Elliott, D. C., Elliott, W. H., and Jones, K. M. (1986). *Data for Biochemical Research,* 3rd ed. (Oxford: Clarendon Press).]

Table A·10 One- and three-letter abbreviations for amino acids

A	Ala	Alanine
B	Asx	Asparagine or aspartate
C	Cys	Cysteine
D	Asp	Aspartate
E	Glu	Glutamate
F	Phe	Phenylalanine
G	Gly	Glycine
H	His	Histidine
I	Ile	Isoleucine
K	Lys	Lysine
L	Leu	Leucine
M	Met	Methionine
N	Asn	Asparagine
P	Pro	Proline
Q	Gln	Glutamine
R	Arg	Arginine
S	Ser	Serine
T	Thr	Threonine
V	Val	Valine
W	Trp	Tryptophan
Y	Tyr	Tyrosine
Z	Glx	Glutamate or glutamine

Table A·11 Abbreviations for some monosaccharides and their derivatives

Monosaccharide or derivative	Abbreviation
Pentoses	
Arabinose	Ara
Ribose	Rib
Xylose	Xyl
Hexoses	
Fructose	Fru
Galactose	Gal
Glucose	Glc
Mannose	Man
Deoxy sugars	
Abequose	Abe
Fucose	Fuc
Rhamnose	Rha
Amino sugars	
Glucosamine	GlcN
Galactosamine	GalN
N-Acetylglucosamine	GlcNAc
N-Acetylgalactosamine	GalNAc
N-Acetylneuraminic acid	NeuNAc
N-Acetylmuramic acid	MurNAc
N-Acetylglucosamine 6-sulfate	GlcNAc-6-SO_4
Sugar acids	
Glucuronic acid	GlcA
Iduronic acid	IdoA

Table A·12 Nomenclature of bases, nucleosides, and nucleotides

Base	Ribonucleoside	Ribonucleotide (5'-monophosphate)
Adenine (A)	Adenosine	Adenosine 5'-monophosphate (AMP); adenylate*
Guanine (G)	Guanosine	Guanosine 5'-monophosphate (GMP); guanylate*
Cytosine (C)	Cytidine	Cytidine 5'-monophosphate (CMP); cytidylate*
Uracil (U)	Uridine	Uridine 5'-monophosphate (UMP); uridylate*

Base	Deoxyribonucleoside	Deoxyribonucleotide (5'-monophosphate)
Adenine (A)	Deoxyadenosine	Deoxyadenosine 5'-monophosphate (dAMP); deoxyadenylate*
Guanine (G)	Deoxyguanosine	Deoxyguanosine 5'-monophosphate (dGMP); deoxyguanylate*
Cytosine (C)	Deoxycytidine	Deoxycytidine 5'-monophosphate (dCMP); deoxycytidylate*
Thymine (T)	Deoxythymidine or thymidine	Deoxythymidine 5'-monophosphate (dTMP); deoxythymidylate* or thymidylate*

*Anionic forms of phosphate esters predominant at pH 7.4

Dictionary of Biochemical Terms

A site. See aminoacyl site

absorptive phase. The period immediately following a meal. During the absorptive phase, which lasts up to four hours in humans, most tissues utilize glucose as fuel.

acceptor stem. The sequence at the 5′ end and the sequence near the 3′ end of a tRNA molecule that are base paired, forming a stem. The acceptor stem is the site of amino acid attachment. Also known as the amino acid stem.

accessory pigments. Pigments other than chlorophyll that are present in photosynthetic membranes. The accessory pigments include carotenoids and phycobilins.

acid. A substance that can donate protons. An acid is converted to its conjugate base by loss of a proton. (The Lewis theory defines an acid as an electron-pair acceptor [Lewis acid].)

acid anhydride. The product formed by condensation of two acids.

acid dissociation constant (K_a). The equilibrium constant for the dissociation of a proton from an acid.

acid protease. See aspartic protease

acid-base catalysis. Catalysis in which the transfer of a proton accelerates a reaction.

acidic solution. A solution that has a pH value less than 7.0.

acidosis. A condition in which the pH of the blood is significantly lower than 7.4.

ACP. See acyl carrier protein

actin filament. A protein filament composed of actin molecules arranged in a twisted, double-stranded rope. Actin filaments are components of the cytoskeletal network and play a role in contractile systems of many organisms. Also known as a microfilament.

activation energy. The free energy required to promote reactants from the ground state to the transition state in a chemical reaction.

activator. See transcriptional activator

active site. The portion of an enzyme that contains the substrate-binding site and the amino acid residues involved in catalyzing the conversion of substrate(s) to product(s). Active sites are usually located in clefts between domains or subunits of proteins or in indentations on the protein surface.

active transport. The process by which a solute specifically binds to a transport protein and is transported across a membrane against the solute concentration gradient. Energy is required to drive active transport. In primary active transport, the energy source may be light, ATP, or electron transport. Secondary active transport is driven by ion concentration gradients.

acyl carrier protein (ACP). A protein (in prokaryotes) or a domain of a protein (in eukaryotes) that binds activated intermediates of fatty acid synthesis via a thioester linkage.

adenosine diphosphate (ADP). A ribonucleoside diphosphate in which two phosphoryl groups are successively linked to the 5′ oxygen atom of adenosine. ADP is formed in reactions in which a phosphoryl group is transferred from adenosine triphosphate (ATP) or to adenosine monophosphate (AMP).

adenosine monophosphate (AMP). A ribonucleoside monophosphate in which a phosphoryl group is linked to the 5′ oxygen atom of adenosine. Phosphorylation of AMP produces adenosine diphosphate (ADP), a precursor of adenosine triphosphate (ATP). Also known as adenylate.

adenosine triphosphate (ATP). A ribonucleoside triphosphate in which three phosphoryl groups are successively linked to the 5′ oxygen atom of adenosine. ATP is the central supplier of energy in living cells, linking exergonic reactions with endergonic reactions. The phosphoanhydride linkages of ATP contain considerable chemical potential energy. By donating phosphoryl groups, ATP can transfer that energy to intermediates that then participate in biosynthetic reactions.

adenylate. See adenosine monophosphate

adipocyte. A triacylglycerol-storage cell found in animals. An adipocyte consists of a fat droplet surrounded by a thin shell of cytosol in which the nucleus and other organelles are suspended.

adipose tissue. Animal tissue composed of specialized triacylglycerol-storage cells known as adipocytes.

A-DNA. The conformation of DNA commonly observed when purified DNA is dehydrated. A-DNA is a right-handed double helix containing approximately 11 base pairs per turn.

ADP. See adenosine diphosphate

aerobic. Occurring in the presence of oxygen.

affinity chromatography. A chromatographic technique used to separate a mixture of proteins or other macromolecules in solution based on specific binding to a ligand that is covalently attached to the chromatographic matrix.

affinity labeling. A process by which an enzyme (or other macromolecule) is covalently tagged by another detectable molecule that specifically interacts with the active site (or other binding site).

aglycone. An organic molecule, such as an alcohol, an amine, or a thiol, that forms a glycosidic linkage with the anomeric carbon of a sugar molecule to form a glycoside.

alarmone. A signal nucleotide that accumulates during metabolic stress in prokaryotes.

aldimine. See Schiff base

aldoses. A class of monosaccharides in which the most oxidized carbon atom, designated C-1, is aldehydic.

alkaline solution. See basic solution

alkalosis. A condition in which the pH of the blood is significantly higher than 7.4.

allosteric effector. See allosteric modulator

allosteric interaction. The modulation of activity of a protein that occurs when a small molecule binds to the regulatory site of the protein.

allosteric modulator. A biomolecule that binds to the regulatory site of an allosteric protein and thereby modulates its activity. An allosteric modulator may be an activator or an inhibitor. Also known as an allosteric effector.

allosteric protein. A protein whose activity is modulated by the binding of another molecule.

allosteric site. See regulatory site

allosteric transitions. The changes in conformation of a protein between the active (R) state and the inactive (T) state, caused by binding or release of an allosteric modulator.

α helix. A common secondary structure of proteins, in which the carbonyl oxygen of each amino acid residue (residue n) forms a hydrogen bond with the amide hydrogen of the fourth residue further toward the C-terminus of the polypeptide chain (residue n + 4). In an ideal right-handed α helix, equivalent positions recur every 0.54 nm, each amino acid residue advances the helix by 0.15 nm along the long axis of the helix, and there are 3.6 amino acid residues per turn.

Ames test. A procedure used to identify potential mutagens. The Ames test measures the ability of a substance to cause reversion of the his⁻ phenotype in a strain of Salmonella typhimurium.

amino acid. An organic acid consisting of an α-carbon atom to which an amino group, a carboxylate group, a hydrogen atom, and a specific side chain (R group) are attached. Amino acids are the building blocks of proteins.

amino acid analysis. A chromatographic procedure used for the separation and quantitation of amino acids in solutions such as protein hydrolysates.

amino acid stem. See acceptor stem

amino terminus. See N-terminus

aminoacyl site. The site on a ribosome that is occupied during protein synthesis by an aminoacyl-tRNA molecule. Also known as the A site.

aminoacyl-tRNA. A tRNA molecule that contains an amino acid covalently attached to the 3′-adenylate residue of the acceptor stem.

aminoacyl-tRNA synthetase. An enzyme that catalyzes the activation and attachment of a specific amino acid to the 3′ end of a corresponding tRNA molecule.

AMP. See adenosine monophosphate

amphibolic pathway. A metabolic pathway that can be both catabolic and anabolic.

amphipathic molecule. A molecule that has both hydrophobic and hydrophilic regions.

anabolic reaction. A metabolic reaction that synthesizes a molecule needed for cell maintenance and growth.

anaerobic. Occurring in the absence of oxygen.

anaplerotic reaction. A reaction that replenishes metabolites removed from a central metabolic pathway.

angstrom (Å). A unit of length equal to 1×10^{-10} m, or 0.1 nm.

anhydride. See acid anhydride

anion. An ion with an overall negative charge.

anode. A positively charged electrode. In electrophoresis, anions move toward the anode.

anomeric carbon. The most oxidized carbon atom of a cyclized monosaccharide. The anomeric carbon has the chemical reactivity of a carbonyl group.

anomers. Isomers of a sugar molecule that have different configurations only at the most oxidized carbon atom.

antenna pigments. Light-absorbing pigments associated with the reaction center of a photosystem. These pigments may form a separate antenna complex or may be bound directly to the reaction-center proteins.

antibiotic. A compound, produced by one organism, that is toxic to other organisms. Clinically useful antibiotics must be specific for pathogens and not affect the human host.

antibody. A glycoprotein synthesized by certain white blood cells as part of the immunological defense system. Antibodies specifically bind to foreign compounds, called antigens, forming antibody-antigen complexes that mark the antigen for destruction. Also known as an immunoglobulin.

anticodon. A sequence of three nucleotides in the anticodon loop of a tRNA molecule. The anticodon binds to the complementary codon in mRNA during translation.

anticodon arm. The stem-and-loop structure in a tRNA molecule that contains the anticodon.

antigen. A molecule specifically bound by an antibody.

antiport. The cotransport of two different species of ions or molecules in opposite directions across a membrane by a transport protein.

antisense RNA. An RNA molecule that binds to a complementary mRNA molecule, forming a double-stranded region that inhibits translation of the mRNA.

apoprotein. A protein whose cofactor(s) is absent. Without the cofactor(s), the apoprotein lacks the biological activity characteristic of the corresponding holoprotein.

aspartic protease. A protease that contains two aspartate residues in the catalytic center, one of which acts as a base catalyst and the other, as an acid catalyst. Aspartic proteases have a pH optimum of about 2–4. Also known as an acid protease.

atomic mass unit. The unit of atomic weight equal to $1/12$th the mass of the ^{12}C isotope of carbon. The mass of the ^{12}C nuclide is exactly 12 by definition.

ATP. See adenosine triphosphate

ATPase. An enzyme that catalyzes hydrolysis of ATP to ADP + P_i. Ion-transporting ATPases use the energy of ATP to transport NaM, KM, or CaH across a cellular membrane.

attenuation. A mechanism of translational regulation in which the rate of ribosome movement along an mRNA molecule determines whether transcription proceeds or terminates. Attenuation allows prokaryotes to regulate expression of an entire operon by the ability to synthesize a short peptide encoded at the beginning of the operon.

autoimmune disease. A human disorder in which the body produces antibodies against normal tissues and cellular components.

autophosphorylation. Phosphorylation of a protein kinase catalyzed by another molecule of the same kinase.

autosome. A chromosome other than a sex chromosome.

autotroph. An organism that can survive using CO_2 as its only source of carbon.

backbone. **1.** The repeating N—C_α—C units connected by peptide bonds in a polypeptide chain. **2.** The repeating sugar-phosphate units connected by phosphodiester linkages in a nucleic acid.

bacteriophage. A virus that infects a bacterial cell.

base. 1. A substance that can accept protons. A base is converted to its conjugate acid by addition of a proton. (The Lewis theory defines a base as an electron-pair donor [Lewis base].) 2. The substituted pyrimidine or purine of a nucleoside or nucleotide. The heterocyclic bases of nucleosides and nucleotides can participate in hydrogen bonding.

base pairing. The interaction between the bases of nucleotides in single-stranded nucleic acids to form double-stranded molecules, such as DNA, or regions of double-stranded secondary structure. The most common base pairs are formed by hydrogen bonding of adenine (A) with thymine (T) or uracil (U) and of guanine (G) with cytosine (C).

basic solution. A solution that has a pH value greater than 7.0. Also known as an alkaline solution.

B-DNA. The most common conformation of DNA and the one proposed by Watson and Crick. B-DNA is a right-handed double helix with a diameter of 2.37 nm and approximately 10.4 base pairs per turn.

β-oxidation pathway. The metabolic pathway that degrades fatty acids to acetyl CoA, producing NADH and QH_2 and thereby generating large amounts of ATP. Each round of β oxidation of fatty acids consists of four steps: oxidation, hydration, further oxidation, and thiolysis.

β sheet. A common secondary structure of proteins that consists of extended polypeptide chains stabilized by hydrogen bonds between the carbonyl oxygen of one peptide bond and the amide hydrogen of another on the same or an adjacent polypeptide chain. The hydrogen bonds are nearly perpendicular to the extended polypeptide chains, which may be either parallel (running in the same N- to C-terminal direction) or antiparallel (running in opposite directions).

β strand. An extended polypeptide chain within a β sheet secondary structure or having the same conformation as a strand within a β sheet.

bile. A suspension of bile salts, bile pigments, and cholesterol that originates in the liver and is stored in the gall bladder. Bile is secreted into the small intestine during digestion.

binary fission. The process by which prokaryotic cells divide. During binary fission, each strand of the double-stranded DNA molecule of the prokaryotic genome directs the synthesis of a complementary strand, the two newly created duplex molecules separate, and the cell divides, with each daughter cell containing one copy of the genetic material.

binding-change mechanism. A proposed mechanism for the formation and release of ATP from F_OF_1 ATP synthase. The mechanism proposes three different binding-site conformations for ATP synthase: an open site from which ATP has been released, an ATP-bearing tight-binding site that is catalytically active, and an ADP and P_i loose-binding site that is catalytically inactive. Inward passage of protons through the ATP synthase complex into the mitochondrial matrix causes the open site to become a loose site; the loose site, already filled with ADP and P_i, to become a tight site; and the ATP-bearing site to become an open site.

bioenergetics. The study of energy changes in biological systems.

biological membrane. See membrane

biopolymer. A biological macromolecule in which many identical or similar small molecules are covalently linked to one another to form a long chain. Proteins, polysaccharides, and nucleic acids are biopolymers.

blunt end. An end of a double-stranded DNA molecule with no single-stranded overhang.

Bohr effect. The phenomenon observed when exposure to carbon dioxide, which lowers the pH inside the cells, causes the oxygen affinity of hemoglobin in red blood cells to decrease.

branch migration. The movement of a crossover, or branch point, resulting in further exchange of DNA strands during recombination.

branch site. The point within an intron that becomes attached to the 5′ end of the intron during mRNA splicing in vertebrates.

buffer. A solution of a weak acid and its conjugate base that resists changes in pH.

buffer capacity. The ability of a solution to resist changes in pH. For a given buffer, maximum buffer capacity is achieved at the pH at which the concentrations of the weak acid and its conjugate base are equal (i.e., when pH = pK_a).

C_3 pathway. See reductive pentose phosphate cycle

C_4 pathway. A pathway for carbon fixation in several plant species that minimizes photorespiration by concentrating CO_2. In this pathway, CO_2 is incorporated into C_4 acids in the mesophyll cells, and the C_4 acids are decarboxylated in the bundle sheath cells, releasing CO_2 for use by the reductive pentose phosphate cycle.

calorie (cal). The amount of energy required to raise the temperature of 1 gram of water by 1°C (from 14.5°C to 15.5°C). One calorie is equal to 4.184 J.

Calvin-Benson cycle. See reductive pentose phosphate cycle

CAM. See Crassulacean acid metabolism

cap. A 7-methylguanosine residue attached by a pyrophosphate linkage to the 5′ end of a eukaryotic mRNA molecule. The cap is added posttranscriptionally and is required for efficient translation. Further covalent modifications yield alternative cap structures.

cap-binding protein (CBP). A eukaryotic translation initiation factor that interacts with the 5′ cap of an mRNA molecule during assembly of the translation initiation complex.

carbanion. A carbon anion that results from the cleavage of a covalent bond between carbon and another atom in which both electrons from the bond remain with the carbon atom.

carbocation. See carbonium ion

carbohydrate. Loosely defined as a compound that is a hydrate of carbon in which the ratio of C:H:O is 1:2:1. Carbohydrates include monomeric sugars (i.e., monosaccharides) and their polymers. Also known as a saccharide.

carbohydrate loading. The athlete's practice of depleting muscle glycogen by intense exercise followed by consumption of high-carbohydrate meals, resulting in higher-than-normal amounts of muscle glycogen.

carbonium ion. A carbon cation that results from the cleavage of a covalent bond between carbon and another atom in which the carbon atom loses both electrons from the bond. Also known as a carbocation.

carboxyl terminus. See C-terminus

carcinogen. An agent that can cause cancer.

carnitine shuttle system. A cyclic pathway that shuttles acetyl CoA from the cytosol to the mitochondria by formation and transport of acyl carnitine.

cascade. Sequential activation of several components, resulting in rapid signal amplification.

catabolic reaction. A metabolic reaction that degrades a complex molecule to provide smaller molecular building blocks and energy to an organism.

catabolite repression. A regulatory mechanism that results in increased rates of transcription of many bacterial genes and operons when glucose is present. A complex between cAMP and cAMP regulatory protein (CRP) activates transcription.

catalytic center. The polar amino acids in the active site of an enzyme that participate in chemical changes during catalysis.

catalytic constant (k_{cat}). A kinetic constant that is a measure of how rapidly an enzyme can catalyze a reaction when saturated with its substrate(s). The catalytic constant is equal to the maximum velocity (V_{max}) divided by the total concentration of enzyme ($[E]_{total}$), or the number of moles of substrate converted to product per mole of enzyme active sites per second, under saturating conditions. Also known as the turnover number.

catalytic triad. The hydrogen-bonded serine, histidine, and aspartate residues in the active site of serine proteases and some other hydrolases. The serine residue is a covalent catalyst; the histidine residue is an acid-base catalyst; and the aspartate residue aligns the histidine residue and stabilizes its protonated form.

cathode. A negatively charged electrode. In electrophoresis, cations move toward the cathode.

cation. An ion with an overall positive charge.

CBP. See cap-binding protein

cDNA. See complementary DNA

cDNA library. A DNA library constructed from cDNA copies of all the mRNA in a given cell type.

cell wall. A mechanically tough, porous outer coat that surrounds the plasma membrane of nearly all bacterial, fungal, and plant cells.

cellulose. A linear (unbranched) homopolymer of glucose residues linked by β-(1→4) glycosidic bonds. A structural polysaccharide of plant cell walls, cellulose accounts for over 50% of the organic matter in the biosphere.

central dogma. The pathway for flow of information from a gene to the corresponding protein. Genetic information is stored in DNA, which can be replicated and passed to daughter cells. Information is copied, or transcribed, from DNA to RNA. RNA is translated during synthesis of a polypeptide chain.

ceramide. A molecule that consists of a fatty acid linked to the C-2 amino group of sphingosine by an amide bond. Ceramides are the metabolic precursors of all sphingolipids.

cerebroside. A glycosphingolipid that contains one monosaccharide residue attached via a β-glycosidic linkage to C-1 of a ceramide. Cerebrosides are abundant in nerve tissue and are found in myelin sheaths.

channel. An integral membrane protein with a central aqueous passage, which allows appropriately sized molecules and ions to traverse the membrane in either direction. Also known as a pore.

channeling. See metabolite channeling

chaotropic agent. A substance that enhances the solubility of nonpolar compounds in water by disrupting regularities in hydrogen bonding among water molecules. Concentrated solutions of chaotropic agents, such as urea and guanidinium salts, decrease the hydrophobic effect and are thus effective protein denaturants.

chaperone. A protein that forms complexes with newly synthesized polypeptide chains and assists in their correct folding into biologically functional conformations. Chaperones may also prevent the formation of incorrectly folded intermediates, prevent incorrect aggregation of unassembled protein subunits, assist in translocation of polypeptide chains across membranes, and assist in the assembly and disassembly of large multiprotein structures.

charge-charge interaction. A noncovalent electrostatic interaction between two charged particles.

chelate effect. The phenomenon by which the constant for binding of a ligand having two or more binding sites to a molecule or atom is greater than the constant for binding of separate ligands to the same molecule or atom.

chemiosmotic theory. A theory proposing that a proton concentration gradient established during oxidation of substrates provides the energy to drive processes such as the formation of ATP from ADP and P_i.

chemotroph. An organism that requires organic compounds other than CO_2 as sources of matter and chemical energy.

chiral atom. An atom with asymmetric substitution that can exist in two different configurations.

chitin. A linear homopolymer of N-acetylglucosamine residues joined by β-(1→4) linkages. Chitin is found in the exoskeletons of insects and crustaceans and in the cell walls of most fungi and many algae and is the second most abundant organic compound on earth.

chlorophyll. A green pigment in photosynthetic membranes that is the principal light-harvesting component in phototrophic organisms.

chloroplast. A chlorophyll-containing organelle in algae and plant cells that is the site of photosynthesis.

chromatin. A DNA-protein complex in the nuclei of eukaryotic cells that is the basic genetic material.

chromatography. A technique used to separate components of a mixture based on their partitioning between a mobile phase, which can be gas or liquid, and a stationary phase, which is a liquid or solid.

chromosome. A single DNA molecule containing many genes. An organism may have a genome consisting of a single chromosome or many.

chromosome walking. A technique for ordering DNA fragments in a genomic library. Chromosome walking involves hybridization, restriction mapping, and isolation of progressively overlapping recombinant DNA molecules.

chylomicron. A type of plasma lipoprotein that transports triacylglycerols, cholesterol, and cholesteryl esters from the small intestine to the tissues.

circular DNA. A DNA molecule in which the two putative ends are covalently linked by $3'$–$5'$ phosphodiester bonds, forming a closed circle.

cisterna. The lumen of a vesicle of the Golgi apparatus.

citrate transport system. A cyclic pathway that shuttles acetyl CoA from the mitochondria to the cytosol, with oxidation of cytosolic NADH to $NAD^{\oplus}$ and reduction of cytosolic $NADP^{\oplus}$ to NADPH. Two molecules of ATP are consumed in each round of the pathway.

citric acid cycle. A metabolic cycle consisting of eight enzyme-catalyzed reactions that completely oxidizes acetyl units to CO_2. The energy released in the oxidation reactions is conserved as reducing power when the coenzymes NAD+ and ubiquinone (Q) are reduced. Oxidation of one molecule of acetyl CoA by the citric acid cycle generates three molecules of NADH, one molecule of QH_2, and one molecule of GTP or ATP. Also known as the Krebs cycle and the tricarboxylic acid cycle.

clone. One of the identical copies derived from the replication or reproduction of a single molecule, cell, or organism.

cloning. The generation of many identical copies of a molecule, cell, or organism. Cloning sometimes refers to the entire process of constructing and propagating a recombinant DNA molecule.

cloning vector. A DNA molecule that carries a segment of foreign DNA. A cloning vector introduces the foreign DNA into a cell where it can be replicated and sometimes expressed.

coding strand. The strand of DNA within a gene whose nucleotide sequence is identical to that of the transcribed RNA (with the replacement of T by U in RNA).

codon. A sequence of three nucleotide residues in mRNA (or DNA) that specifies a particular amino acid according to the genetic code.

coenzyme. An organic molecule required by an enzyme for full activity. Coenzymes can be further classified as cosubstrates or prosthetic groups.

cofactor. An inorganic ion or organic molecule required by an apoenzyme to convert it to a holoenzyme. There are two types of cofactors: essential ions and coenzymes.

competitive inhibition. Inhibition of an enzyme-catalyzed reaction by an inhibitor that prevents substrate binding.

complementary DNA (cDNA). Double-stranded DNA synthesized from an mRNA template by the action of reverse transcriptase followed by DNA polymerase.

complementation. A technique for selecting transformed cells by testing for the ability of the recombinant DNA to supply a gene product missing in the host cell.

concerted theory of cooperativity and allosteric regulation. A model of the cooperative binding of identical ligands to oligomeric proteins. According to the simplest form of the concerted theory, the change in conformation of a protein due to the binding of a substrate or an allosteric modulator shifts the equilibrium of the conformation of the protein between T (a low substrate-affinity conformation) and R (a high substrate-affinity conformation). This theory suggests that all subunits of the protein have the same conformation, either all T or all R. Also known as the symmetry-driven theory.

condensation. A reaction involving the joining of two or more molecules accompanied by the elimination of water, alcohol, or other simple substance.

configuration. A spatial arrangement of atoms, which cannot be altered without breaking and re-forming covalent bonds.

conformation. Any three-dimensional structure, or spatial arrangement, of a molecule that results from rotation of functional groups around single bonds. Because there is free rotation around single bonds, a molecule can potentially assume many conformations.

conjugate acid. The product resulting from the gain of a proton by a base.

conjugate base. The product resulting from the loss of a proton by an acid.

conjugation. The passage of genetic material from one bacterium to another through the sex pilus.

consensus sequence. The sequence of nucleotides most commonly found at each position within a region of DNA or RNA.

constitutive gene expression. Expression of a gene at a steady level. Housekeeping genes are often expressed constitutively.

converter enzyme. An enzyme that catalyzes the covalent modification of another enzyme, thereby changing its catalytic activity.

cooperativity. 1. The phenomenon whereby the binding of one ligand or substrate molecule to a protein influences the affinity of the protein for additional molecules of the same substance. Cooperativity may be positive or negative. 2. The phenomenon whereby formation of structure in one part of a macromolecule promotes the formation of structure in the rest of the molecule.

core particle. See nucleosome core particle

corepressor. A ligand that binds to and activates a repressor of a gene.

Cori cycle. An interorgan metabolic loop that recycles carbon and transports energy from the liver to the peripheral tissues. Glucose is released from the liver and metabolized to produce ATP in other tissues. The resulting lactate is then returned to the liver for conversion back to glucose by gluconeogenesis.

cosmid. A cloning vector that accommodates large fragments of insert DNA. Cosmids allow efficient transfection but permit propagation of recombinant DNA molecules as plasmids.

cosubstrate. A coenzyme that is a substrate in an enzyme-catalyzed reaction. A cosubstrate is altered during the course of the reaction and dissociates from the active site of the enzyme. The original form of the cosubstrate can be regenerated in a subsequent enzyme-catalyzed reaction.

cotranscriptional processing. RNA processing that occurs before transcription is complete.

cotranslational modification. Covalent modification of a protein that occurs before elongation of the polypeptide is complete.

cotransport. The coupled transport of two different species of solutes across a membrane, in the same direction (symport) or the opposite direction (antiport), carried out by a transport protein.

coupled reactions. Two metabolic reactions that share a common intermediate.

covalent catalysis. Catalysis in which one substrate, or part of it, forms a covalent bond with the catalyst and then is transferred to a second substrate. Many enzymatic group-transfer reactions proceed by covalent catalysis.

Crassulacean acid metabolism (CAM). A modified sequence of carbon-assimilation reactions used primarily by plants in arid environments to reduce water loss during photosynthesis. In these reactions, CO_2 is taken up at night, resulting in the formation of malate. During the day, malate is decarboxylated, releasing CO_2 for use by the reductive pentose phosphate cycle.

cristae. The folds of the inner mitochondrial membrane.

cruciform structure. The crosslike conformation adopted by double-stranded DNA when inverted repeats form a base-paired structure involving the complementary regions within the same strand.

C-terminus. The amino acid residue bearing a free a-carboxyl group at one end of a peptide chain. Also known as the carboxyl terminus.

cumulative feedback inhibition. Inhibition of an enzyme that catalyzes an early step in several biosynthetic pathways by intermediates or end products of those pathways. The degree of inhibition increases as more of the inhibitors bind.

cyclic electron transport. A modified sequence of electron-transport steps in chloroplasts that operates to provide ATP without the simultaneous formation of NADPH.

cytochrome. A heme-containing protein that is an electron carrier in processes such as respiration and photosynthesis.

cytoplasm. The part of a cell enclosed by the plasma membrane, excluding the nucleus.

cytoskeleton. A network of proteins that contributes to the structure and organization of a eukaryotic cell.

cytosol. The aqueous portion of the cytoplasm minus the subcellular structures.

D arm. The stem-and-loop structure in a tRNA molecule that contains dihydrouridylate (D) residues.

dalton. A unit of mass equal to one atomic mass unit.

dark reactions. The photosynthetic reactions in which NADPH and ATP are used to reduce CO_2 to carbohydrate. Also known as the light-independent reactions.

de novo pathway. A metabolic pathway in which a biomolecule is formed from simple precursor molecules.

deaminase. An enzyme that catalyzes the removal of an amino group from a substrate, releasing ammonia.

degeneracy. The existence of several different codons that specify the same amino acid.

dehydrogenase. An enzyme that catalyzes the removal of hydrogen from a substrate or the oxidation of a substrate. Dehydrogenases are members of the IUB class of enzymes known as oxidoreductases.

denaturation. 1. A disruption in the native conformation of a biological macromolecule that results in loss of the biological activity of the macromolecule. 2. The complete unwinding and separation of complementary strands of DNA.

deoxyribonuclease (DNase). An enzyme that catalyzes the hydrolysis of deoxyribonucleic acids to form oligodeoxynucleotides and/or monodeoxynucleotides.

deoxyribonucleic acid (DNA). A polymer consisting of deoxyribonucleotide residues joined by 3´-5´ phosphodiester bonds. The sugar moiety in DNA is 2-deoxyribose. DNA is the genetic material of all cells and many viruses.

detergent. An amphipathic molecule consisting of a hydrophobic portion and a hydrophilic end that may be ionic or polar. Detergent molecules can aggregate in aqueous media to form micelles. Also known as a surfactant.

diabetes mellitus. A metabolic disease characterized by hyperglycemia resulting from abnormal regulation of fuel metabolism by insulin. Diabetes mellitus may arise from the lack of insulin or from poor responsiveness of cells to insulin.

dialysis. A procedure in which low-molecular-weight solutes in a sample are removed by diffusion through a semipermeable barrier and replaced by solutes from the surrounding medium.

diffusion-controlled reaction. A reaction that occurs with every collision between reactant molecules. In enzyme-catalyzed reactions, the k_{cat}/K_m ratio approaches a value of 10^8–$10^9 M^{-1} s^{-1}$.

digestion. The process by which dietary macromolecules are hydrolyzed to smaller molecules that can be absorbed by an organism.

diploid. Having two sets of chromosomes or two copies of the genome.

dipole. Two equal but opposite charges, separated in space, resulting from the uneven distribution of charge within a molecule or a chemical bond.

direct repair. The removal of DNA damage by proteins that recognize damaged nucleotides and mismatched bases and repair them without cleaving the DNA or excising the base.

distributive enzyme. An enzyme that dissociates from its growing polymeric product after addition of each monomeric unit and must reassociate with the polymer for polymerization to proceed.

disulfide bond. A covalent linkage formed by oxidation of two thiol groups. Disulfide bonds are important in stabilizing the three-dimensional structures of some proteins.

DNA. See deoxyribonucleic acid

DNA fingerprinting. Analysis of the genetic polymorphism of different individuals.

DNA gyrase. See topoisomerase

DNA library. The set of recombinant DNA molecules generated by ligating all the fragments of a sample of DNA into vectors.

DNA ligase. The enzyme that joins two DNA polynucleotides by catalyzing the formation of a phosphodiester bond. DNA ligase can also repair gaps in double-stranded DNA.

DNA polymerase. An enzyme that catalyzes the DNA template–directed addition of nucleotide residues to the 3´ end of an existing polynucleotide. Some DNA polymerases contain exonuclease activity used in proofreading newly polymerized sequences.

DNase. See deoxyribonuclease

domain. A discrete, independent folding unit within the tertiary structure of a protein. Domains are usually combinations of several motifs.

dosage compensation. Selective inactivation of certain genes present in multiple copies within the cell to control the level of gene expression.

double helix. A nucleic acid conformation in which two antiparallel polynucleotide strands wrap around each other to form a two-stranded helical structure stabilized largely by stacking interactions between adjacent hydrogen-bonded base pairs.

double-reciprocal plot. A plot of the reciprocal of initial velocity versus the reciprocal of substrate concentration for an enzyme-catalyzed reaction. The x and y intercepts indicate the values of the reciprocals of the Michaelis constant and the maximum velocity, respectively. A double-reciprocal plot is a linear transformation of the Michaelis-Menten equation. Also known as a Lineweaver-Burk plot.

E. See reduction potential

E°′. See standard reduction potential

E site. See exit site

Edman degradation. A procedure used to determine the sequence of amino acid residues from a free N-terminus of a polypeptide chain. The N-terminal residue is chemically modified, cleaved from the chain, and identified by chromatographic procedures, and the rest of the polypeptide is recovered. Multiple reaction cycles allow identification of the new N-terminal residue generated by each cleavage step.

effective molarity. An expression of the catalytic power of an enzyme in terms of the rate of a nonenzymatic reaction within a single compound relative to the rate of a reaction of two compounds possessing the same reactive groups as the single compound.

effector enzyme. A membrane-associated protein that produces an intracellular second messenger in response to a signal from a transducer.

eicosanoid. An oxygenated derivative of a 20-carbon polyunsaturated fatty acid. Eicosanoids function as short-range messengers in the regulation of various physiological processes.

electrical potential. See membrane potential

electrogenic transport. The transfer of ionic solutes across a membrane by a process that results in a net transfer of charge, thereby creating changes in membrane potential.

electromotive force (emf). A measure of the difference between the reduction potentials of the reactions on the two sides of an electrochemical cell (i.e., the voltage difference produced by the reactions).

electroneutral transport. The transfer of ionic solutes across a membrane by a process that results in no net transfer of charge.

electron-transport chain. See respiratory electron-transport chain

electrophile. A positively charged or electron-deficient species that is attracted to chemical species that are negatively charged or contain unshared electron pairs (nucleophiles).

electrophoresis. A technique used to separate molecules by their migration in an electric field, primarily on the basis of their net charge.

electrostatic interaction. A general term for the electronic interaction between particles. Electrostatic interactions include charge-charge interactions, hydrogen bonds, and van der Waals forces.

elongation factor. A protein that is involved in extending the peptide chain during protein synthesis.

emf. See electromotive force

enantiomers. Stereoisomers that are nonsuperimposable mirror images.

endergonic reaction. A chemical reaction that is characterized by a positive free-energy change. Such a reaction cannot occur spontaneously and requires the input of energy from outside the system to proceed.

endocytosis. The process by which matter is engulfed by a plasma membrane and brought into the cell within a lipid vesicle derived from the membrane.

endonuclease. An enzyme that catalyzes the hydrolysis of phosphodiester linkages at various sites within polynucleotide chains.

endoplasmic reticulum. A membranous network of tubules and sheets continuous with the outer nuclear membrane of eukaryotic cells. Regions of the endoplasmic reticulum coated with ribosomes are called the rough endoplasmic reticulum; regions having no attached ribosomes are known as the smooth endoplasmic reticulum. The endoplasmic reticulum is involved in the synthesis, sorting, and transport of certain proteins and in the synthesis of lipids.

endosomes. Smooth vesicles inside the cell that are receptacles for endocytosed material.

endotoxin. See lipopolysaccharide

energy-rich compound. A compound whose hydrolysis occurs with a large negative free-energy change (greater than that for $ATP \rightarrow ADP + P_i$). The free energy available upon cleavage of an energy-rich compound can be used to drive other reactions.

enhancer. A region of DNA, located some distance from the promoter, to which a transcriptional activator binds, thereby increasing the rate of transcription.

enteric bacteria. The bacteria that inhabit the large intestine of animals. Most enteric bacteria are facultative anaerobes.

enthalpy (H). A thermodynamic state function that describes the heat content of a system.

entropy (S). A thermodynamic state function that describes the randomness or disorder of a system.

enzymatic reaction. A reaction catalyzed by a biological catalyst, an enzyme. Enzymatic reactions are 10^3 to 10^{17} times faster than the corresponding uncatalyzed reactions.

enzyme. A biological catalyst, almost always a protein. Some enzymes may require additional cofactors for activity. Virtually all biochemical reactions are catalyzed by specific enzymes.

enzyme assay. A method used to analyze the activity of a sample of an enzyme. Typically, enzymatic activity is measured under selected conditions such that the rate of conversion of substrate to product is proportional to enzyme concentration.

enzyme inhibitor. A compound that binds to an enzyme and interferes with its activity by preventing either the formation of the ES complex or its conversion to E + P.

enzyme-substrate complex (ES). A complex formed when substrate molecules bind noncovalently within the active site of an enzyme.

epimers. Isomers that differ in configuration at only one of several chiral centers.

equilibrium. The state of a system in which the rate of conversion of substrate to product is equal to the rate of conversion of product to substrate. The free-energy change for a reaction or system at equilibrium is zero.

equilibrium constant (K_{eq}). The ratio of the concentrations of products to the concentrations of reactants at equilibrium. The equilibrium constant is related to the standard free-energy change of a reaction.

equilibrium density gradient centrifugation. A technique used to separate macromolecules of different densities in an ultracentrifuge based on their buoyancy in an appropriate density gradient.

error-prone repair. The response to DNA damage that cannot be bypassed by the replication machinery. In error-prone repair, RecA protein binds to single-stranded gaps in DNA near the replication fork. Additional proteins allow the replication complex to proceed through the damaged region, haphazardly inserting nucleotides opposite unrecognizable template-strand nucleotides.

escape synthesis. Low-level transcription of a repressed gene that occurs even in the absence of induction. Escape synthesis results from spontaneous dissociation of the repressor for brief interludes.

essential amino acid. An amino acid that cannot be synthesized by an animal and must be obtained in the diet.

essential fatty acid. A fatty acid that cannot be synthesized by an animal and must be obtained in the diet.

essential ion. An ion required as a cofactor for the catalytic activity of certain enzymes. Some essential ions, called activator ions, are reversibly bound to enzymes and often participate in the binding of substrates, whereas tightly bound metal ions frequently participate directly in catalytic reactions.

eukaryote. An organism whose cells generally possess a nucleus and internal membranes.

excision repair. The reversal of DNA damage by excision-repair endonucleases. Gross lesions that alter the structure of the DNA helix are repaired by cleavage on each side of the lesion and removal of the damaged DNA. The resulting single-stranded gap is filled by DNA polymerase and sealed by DNA ligase.

exergonic reaction. A chemical reaction that is characterized by a negative free-energy change. Such a reaction is spontaneous in that it does not require the input of additional energy to proceed.

exit site. The site on a ribosome from which a deaminoacylated tRNA molecule is released during protein synthesis. Also known as the E site.

exocytosis. The process by which material destined for secretion from a cell is enclosed in lipid vesicles that are transported to and fuse with the plasma membrane, releasing the material into the extracellular space.

exon. A nucleotide sequence that is present in the primary RNA transcript and in the mature RNA molecule. The term exon also refers to the region of the gene that encodes the corresponding RNA exon.

exonuclease. An enzyme that catalyzes the sequential hydrolysis of phosphodiester linkages from one end of a polynucleotide chain.

expression vector. A cloning vector that allows inserted DNA to be transcribed and translated into protein.

extrinsic membrane protein. See peripheral membrane protein

facilitated diffusion. See passive transport

facultative anaerobe. An organism that can survive in the presence or absence of oxygen.

fat-soluble vitamin. See lipid vitamin

fatty acid. A long-chain aliphatic hydrocarbon with a single carboxyl group at one end. Fatty acids are the simplest type of lipid and are components of many more complex lipids, including triacylglycerols, glycerophospholipids, sphingolipids, and waxes.

feedback inhibition. Inhibition of an enzyme that catalyzes an early step in a metabolic pathway by an end product of the same pathway.

feed-forward activation. Activation of an enzyme in a metabolic pathway by a metabolite produced earlier in the pathway.

fermentation. The anaerobic catabolism of metabolites for energy production. In alcoholic fermentation, pyruvate is converted to ethanol and carbon dioxide.

fibrous proteins. A major class of water-insoluble proteins that are often built upon a single repetitive structure. Many fibrous proteins are physically tough and provide mechanical support to individual cells or entire organisms.

fingerprinting. See DNA fingerprinting

first-order reaction. A reaction whose rate is directly proportional to the concentration of only one reactant.

Fischer projection. A two-dimensional representation of the three-dimensional structures of sugars and related compounds. In a Fischer projection, the carbon skeleton is drawn vertically, with C-1 at the top. At a chiral center, horizontal bonds extend toward the viewer and vertical bonds extend away from the viewer.

fluid mosaic model. A model proposed for the structure of biological membranes. In this model, the membrane is depicted as a dynamic structure in which lipids and membrane proteins (both integral and peripheral) rotate and undergo lateral diffusion.

fluorescence. A form of luminescence in which visible radiation is emitted from a molecule as it passes from a higher to a lower electronic state.

flux. The flow of material through a metabolic pathway. Flux depends on the supply of substrates, the removal of products, and the catalytic capabilities of the enzymes involved in the pathway.

footprinting. A technique used to identify the sequence of DNA or RNA bound by a protein. The bound protein protects specific nucleotide sequences from chemical or enzymatic digestion. The protected regions are visualized as gaps, or footprints, when the ladder of nucleic acid fragments is resolved by gel electrophoresis. Also known as nuclease protection.

frameshift mutation. An alteration in DNA caused by the insertion or deletion of a number of nucleotides not divisible by three. A frameshift mutation changes the reading frame of the corresponding mRNA molecule and affects translation of all codons downstream of the mutation.

free radical. A molecule or atom with an unpaired electron.

free-energy change (ΔG). A thermodynamic quantity that defines the equilibrium condition in terms of the changes in enthalpy (H) and entropy (S) of a system at constant pressure. $\Delta G = \Delta H - T\Delta S$, where T is absolute temperature. Free energy is a measure of the energy available within a system to do work.

freeze-fracture electron microscopy. A technique used to visualize the structure of biological membranes. During freeze-fracture, the membrane sample is rapidly frozen and then split along the interface between the leaflets of the lipid bilayer. The exposed membrane surface is then coated with a thin metal film, producing a replica of the leaflet surface which can be visualized using an electron microscope.

furanose. A monosaccharide structure that forms a five-membered ring as a result of intramolecular hemiacetal formation.

G protein. A protein that binds guanine nucleotides.

ΔG. See free-energy change

ΔG°′. See standard free-energy change

ganglioside. A glycosphingolipid in which oligosaccharide chains containing N-acetylneuraminic acid are attached to a ceramide. Gangliosides are present on cell surfaces and provide cells with distinguishing surface markers that may serve in cellular recognition and cell-to-cell communication.

gas chromatography. A chromatographic technique used to separate components of a mixture based on their partitioning between the gas phase and a stationary phase, which can be a liquid or solid.

gastrointestinal tract. The digestive system of an animal. In the gastrointestinal tract, dietary biopolymers, such as starch and proteins, are enzymatically hydrolyzed to their monomeric units, and the monomers are absorbed.

gel-filtration chromatography. A chromatographic technique used to separate a mixture of proteins or other macromolecules in solution based on molecular size, using a matrix of porous beads. Also known as molecular-exclusion chromatography.

gene. Loosely defined as a segment of DNA that is transcribed. In some cases, the term gene may also be used to refer to a segment of DNA that encodes a functional protein or RNA molecule.

gene amplification. Regulation of gene expression by increasing the number of copies of a gene within a cell.

gene replacement therapy. A technique in which cells are removed from a patient suffering from an enzyme deficiency, the gene for wild-type enzyme is transferred into the cells in culture, and the cells are returned to the patient.

general excision-repair pathway. See excision repair

genetic code. The correspondence between the sequence of residues in a nucleic acid and the amino acid sequence in a protein. A sequence of three nucleotide residues, known as a codon, specifies a single amino acid. The standard genetic code, composed of 64 codons, is used by almost all living organisms.

genetic engineering. See recombinant DNA technology

genetic load. The overall risk of harmful or lethal mutation. The genetic load is proportional to the size of the genome of the organism.

genetic recombination. The exchange or transfer of DNA from one chromosome to another. Homologous recombination is the exchange or transfer of DNA between molecules with very similar or identical nucleotide sequences.

genetic transformation. A process by which a cell takes up intact DNA from outside the cell.

genome. The total amount of genetic information in an organism. It is equivalent to a single complete chromosome or to a set of chromosomes (haploid). Mitochondria and chloroplasts have genomes separate from that in the nucleus of eukaryotic cells.

genomic library. A DNA library constructed by fragmenting and cloning all the DNA for the genome of an organism.

globular proteins. A major class of proteins, many of which are water soluble. Globular proteins are compact and roughly spherical, containing tightly folded polypeptide chains. Typically, globular proteins include indentations, or clefts, which specifically recognize and transiently bind other compounds.

glucogenic compound. A compound, such as an amino acid, that can be used for gluconeogenesis in animals.

gluconeogenesis. A pathway for synthesis of glucose from a noncarbohydrate precursor. Gluconeogenesis from pyruvate involves the seven near-equilibrium reactions of glycolysis traversed in the reverse direction. The three metabolically irreversible reactions of glycolysis are bypassed by four enzymatic reactions that do not occur in glycolysis.

glucose homeostasis. The maintenance of constant levels of glucose in the circulation achieved by balancing glucose synthesis or absorption against utilization.

glucose-alanine cycle. An interorgan metabolic loop that transports nitrogen to the liver and transports energy from the liver to the peripheral tissues. Glucose is released from the liver and metabolized in muscle to pyruvate, with concomitant production of ATP. Pyruvate can be converted to alanine in muscle. Alanine is returned to the liver, where it is metabolized to ammonia and pyruvate. The ammonia is incorporated into urea, and the pyruvate is converted back to glucose by gluconeogenesis.

glucose–fatty acid cycle. An interorgan regulatory system that operates to provide fatty acids as an alternative fuel to glucose. When glucose levels are low, insulin levels are low, and free fatty acids are released from adipocytes. The metabolism of fatty acids generates inhibitors of glucose metabolism, thus sparing the use of glucose.

glycan. A general term for an oligosaccharide or a polysaccharide. A homoglycan is a polymer of identical monosaccharide residues; a heteroglycan is a polymer of different monosaccharide residues.

glycerophospholipid. A lipid consisting of two fatty acyl groups bound to C-1 and C-2 of glycerol 3-phosphate and, in most cases, a polar substituent attached to the phosphate moiety. Glycerophospholipids are major components of biological membranes.

glycocalyx. A coat of lipid- and protein-bound carbohydrate that extends from the extracellular surface of the eukaryotic plasma membrane.

glycoconjugate. A carbohydrate derivative in which one or more carbohydrate chains are covalently linked to a peptide chain, protein, or lipid.

glycoforms. Glycoproteins containing identical amino acid sequences but different oligosaccharide-chain compositions.

glycogen. A branched homopolymer of glucose residues joined by α-(1→4) linkages with α-(1→6) linkages at branch points. Glycogen is a storage polysaccharide in animals and bacteria.

glycogenolysis. The pathway for intracellular degradation of glycogen.

glycolysis. A catabolic pathway consisting of 10 enzyme-catalyzed reactions by which one molecule of glucose is converted to two molecules of pyruvate. In the process, two molecules of ATP are formed from ADP + P_i, and two molecules of $NAD^{\oplus}$ are reduced to NADH.

glycoprotein. A protein that contains covalently bound carbohydrate residues.

glycosaminoglycan. An unbranched polysaccharide of repeating disaccharide units. One component of the disaccharide is an amino sugar; the other component is usually a uronic acid.

glycoside. A molecule containing a carbohydrate in which the hydroxyl group of the anomeric carbon has been replaced through condensation with an alcohol, an amine, or a thiol.

glycosidic bond. Acetal linkage formed by condensation of the anomeric carbon atom of a saccharide with a hydroxyl, amino, or thiol group of another molecule. The most commonly encountered glycosidic bonds are formed between the anomeric carbon of one sugar and a hydroxyl group of another sugar. Nucleosidic bonds are N-linked glycosidic bonds.

glycosphingolipid. A lipid containing sphingosine and carbohydrate moieties.

glycosylation. See protein glycosylation.

glyoxylate cycle. A variation of the citric acid cycle in certain plants, bacteria, and yeast that allows net production of glucose from acetyl CoA via oxaloacetate. The glyoxylate cycle bypasses the two CO_2-producing steps of the citric acid cycle.

glyoxysome. An organelles that contains specialized enzymes for the glyoxylate cycle.

Golgi apparatus. A complex of flattened, fluid-filled membranous sacs in eukaryotic cells, often found in proximity to the endoplasmic reticulum. The Golgi apparatus is involved in the modification, sorting, and targeting of proteins.

GPI membrane anchor. A phosphatidylinositol-glycan structure attached to a protein via a phosphoethanolamine residue. The fatty acyl moieties of the GPI anchor are embedded in the lipid bilayer, thereby tethering the protein to the membrane.

granal lamellae. Regions of the thylakoid membrane that are located within grana and are not in contact with the stroma.

granum. A stack of flattened vesicles formed from the thylakoid membrane in chloroplasts.

group translocation. A type of primary active transport during which the translocated species is chemically modified.

group-transfer potential. See phosphoryl-group–transfer potential

group-transfer protein. See protein coenzyme

group-transfer reaction. A reaction in which a substituent or functional group is transferred from one substrate to another.

growth factor. A protein that regulates cell proliferation by stimulating resting cells to undergo cell division.

H. See enthalpy

hairpin. 1. A secondary structure adopted by single-stranded polynucleotides that arises when short regions fold back on themselves and hydrogen bonds form between complementary bases. Also known as a stem-loop. 2. A tight turn connecting two consecutive β strands of a polypeptide.

haploid. Having one set of chromosomes or one copy of the genome.

Haworth projection. A representation in which a cyclic sugar molecule is depicted as a flat ring that is projected perpendicular to the plane of the page. Heavy lines represent the part of the molecule that extends toward the viewer.

HDL. See high density lipoprotein

heat of vaporization. The amount of heat required to evaporate 1 gram of a liquid.

heat-shock gene. A gene whose transcription is increased in response to stresses such as high temperature. Many heat-shock genes that encode chaperones are also expressed in the absence of stress.

helicase. An enzyme that is involved in unwinding DNA.

helix-destabilizing protein. See single-strand binding protein

hemiacetal. The product formed when an alcohol reacts with an aldehyde.

hemiketal. The product formed when an alcohol reacts with a ketone.

hemoglobin. A tetrameric, heme-containing globular protein in red blood cells that carries oxygen (O_2) to other cells and tissues.

Henderson-Hasselbalch equation. An equation that describes the pH of a solution in terms of the pK_a of a weak acid and the concentrations of the acid and its conjugate base.

heterochromatin. Regions of chromatin that are highly condensed. The genes in heterochromatin are inaccessible to transcription factors and are not transcribed.

heterocyclic molecule. A molecule that contains a ring structure made up of more than one type of atom.

heterogeneous nuclear RNA (hnRNA). See mRNA precursor

heterotroph. An organism that requires at least one organic nutrient, such as glucose, as a carbon source.

hexose monophosphate shunt. See pentose phosphate pathway

high density lipoprotein (HDL). A type of plasma lipoprotein that is enriched in protein and transports cholesterol and cholesteryl esters from tissues to the liver.

high-pressure liquid chromatography (HPLC). A chromatographic technique used to separate components of a mixture by dissolving the mixture in a liquid solvent and forcing it to flow through a chromatographic column under high pressure.

histones. A class of proteins that bind to DNA to form chromatin. The nuclei of most eukaryotic cells contain five histones, known as H1, H2A, H2B, H3, and H4.

hnRNA. See mRNA precursor

Holliday junction. The region of strand crossover resulting from recombination between two molecules of homologous double-stranded DNA.

holoprotein. An active protein possessing all of its cofactors and subunits.

homeostasis. The maintenance of a constant metabolic state.

homologous recombination. See genetic recombination

homology. The similarity of genes or proteins as a result of evolution from a common ancestor.

hormone. A regulatory molecule synthesized in one cell or endocrine gland and transported to another cell or tissue. The target cell or tissue contains the appropriate hormone receptor and responds to the extracellular hormone, or first messenger, by producing intracellular second messengers, which alter the activity of certain enzymes.

hormone-response element. A DNA sequence that binds a transcriptional activator consisting of a steroid hormone–receptor complex.

housekeeping genes. Genes that encode proteins or RNA molecules that are essential for the normal activities of all living cells.

HPLC. See high-pressure liquid chromatography

hydration. A state in which a molecule or ion is surrounded by water.

hydrogen bond. A weak electrostatic interaction formed when a hydrogen atom bonded covalently to a strongly electronegative atom also bonds to the unshared electron pair of another electronegative atom.

hydrolase. An enzyme that catalyzes the hydrolytic cleavage of its substrate(s) (i.e., hydrolysis).

hydrolysis. Cleavage of a bond within a molecule by group transfer to water.

hydropathy. A measure of the hydrophobicity of amino acid side chains. The more positive the hydropathy value, the greater the hydrophobicity.

hydrophilicity. The degree to which a compound or functional group interacts with water or is preferentially soluble in water.

hydrophobic effect. The exclusion of hydrophobic groups or molecules by water. The hydrophobic effect appears to depend on the increase in entropy of solvent water molecules that are released from an ordered arrangement around the hydrophobic group.

hydrophobic interaction. A weak, noncovalent interaction between nonpolar molecules or substituents that results from the strong association of water molecules with one another. Such association leads to the shielding or exclusion of nonpolar molecules from an aqueous environment.

hydrophobicity. The degree to which a compound or functional group that is soluble in nonpolar solvents is insoluble or only sparingly soluble in water.

IDL. See intermediate density lipoprotein

immunofluorescence microscopy. A technique for visualizing cellular components by first labeling them with specific fluorescent ligands or antibodies.

immunoglobulin. See antibody

in vitro. Occurring under artificial conditions, such as those in a laboratory, rather than under physiological conditions or in an intact organism.

in vivo. Occurring within a living cell or organism.

induced fit. A model for activation of an enzyme by a substrate-initiated conformational change.

inducer. A ligand that binds to and inactivates a repressor, thereby increasing the transcription of the gene controlled by the repressor.

inhibition constant (K_i). The equilibrium constant for the dissociation of an inhibitor from an enzyme-inhibitor complex.

initial velocity (v_0). The rate of conversion of substrate to product in the early stages of an enzymatic reaction, before appreciable product has been formed.

initiation codon. A codon that specifies the initiation site for protein synthesis. The methionine codon (AUG) is the most common initiation codon.

initiation factor. See translation initiation factor

initiator-tRNA. The tRNA molecule that is used exclusively at initiation codons. In eukaryotes and archaebacteria, the initiator-tRNA is usually a methionyl-tRNA; in eubacteria, the methionine moiety is formylated.

insert DNA. The DNA fragment that is carried by a cloning vector.

insertional inactivation. The inactivation of a gene caused by inserting a DNA fragment into the coding region of the gene.

integral membrane protein. A membrane protein that penetrates the hydrophobic core of the lipid bilayer and usually spans the bilayer completely. Also known as an intrinsic membrane protein.

integron. See transposon

intercalating agent. A compound containing a planar ring structure that can fit between the stacked base pairs of DNA. Intercalating agents distort the DNA structure, partially unwinding the double helix.

interconvertible enzyme. An enzyme whose activity is regulated by covalent modification. Interconvertible enzymes undergo transitions between active and inactive states but may be frozen in one state or the other by a covalent substitution.

intermediary metabolism. The metabolic reactions by which the small molecules of cells are interconverted.

intermediate density lipoprotein (IDL). A type of plasma lipoprotein that is formed during the breakdown of VLDLs.

intermediate filament. A double- or triple-stranded structure composed of different protein subunits, found in the cytoplasm of most eukaryotic cells. Intermediate filaments are components of the cytoskeletal network.

intrinsic membrane protein. See integral membrane protein

intron. An internal nucleotide sequence that is removed from the primary RNA transcript during processing. The term intron also refers to the region of the gene that encodes the corresponding RNA intron.

inverted repeat. A sequence of nucleotides that is repeated in the opposite orientation within the same polynucleotide strand. An inverted repeat in double-stranded DNA can give rise to a cruciform structure.

ion pair. An electrostatic interaction between ionic groups of opposite charge within the interior of a macromolecule such as a globular protein.

ion-product constant for water (K_w). The product of the concentrations of hydronium ions and hydroxide ions in an aqueous solution, equal to 1.0×10^{-14} M^2.

ion-exchange chromatography. A chromatographic technique used to separate a mixture of ionic species in solution, using a charged matrix. In anion-exchange chromatography, a positively charged matrix binds negatively charged solutes, and in cation-exchange chromatography, a negatively charged matrix binds positively charged solutes. The bound species can be serially eluted from the matrix by gradually changing the pH or increasing the salt concentration in the solvent.

ionophore. A compound that facilitates the diffusion of ions across bilayers and membranes by serving as a mobile ion carrier or by forming a channel for ion passage.

isoacceptor tRNA molecules. Different tRNA molecules that bind the same amino acid.

isoelectric focusing. A modified form of electrophoresis that uses buffers to create a pH gradient within a polyacrylamide gel. Each protein migrates to its isoelectric point (pI), that is, the pH in the gradient at which it no longer carries a net positive or negative charge.

isoelectric point (pI). The pH at which a zwitterionic molecule does not migrate in an electric field because its net charge is zero.

isoenzymes. See isozymes

isomerase. An enzyme that catalyzes an isomerization reaction, a change in geometry or structure within one molecule.

isopeptide bond. A peptide bond between the α-carboxylate group of an amino acid and the e-amino group of a lysine residue.

isoprene. A branched, unsaturated five-carbon molecule that forms the basic structural unit of all isoprenoids, including the steroids and lipid vitamins.

isoprenoid. A lipid that is structurally related to isoprene.

isozymes. Different proteins from a single biological species that catalyze the same reaction. Also known as isoenzymes.

junk DNA. Regions of the genome with no apparent function.

K_a. See acid dissociation constant

karyotype. A set of chromosomes visualized by staining.

kb. See kilobase pair

k_{cat}. See catalytic constant

k_{cat}/K_m. The second-order rate constant for conversion of enzyme and substrate to enzyme and product at low substrate concentrations. The ratio of kcat to Km, when used to compare several substrates, is called the specificity constant.

K_{eq}. See equilibrium constant

ketimine. See Schiff base

ketogenesis. The pathway that synthesizes ketone bodies from acetyl CoA in the mitochondrial matrix in mammals.

ketogenic compound. A compound, such as an amino acid, that can be degraded to form acetyl CoA and can thereby contribute to the synthesis of fatty acids or ketone bodies.

ketone bodies. Small molecules that are synthesized in the liver from acetyl CoA. During starvation, the ketone bodies β-hydroxybutyrate and acetoacetate become major metabolic fuels.

ketoses. A class of monosaccharides in which the most oxidized carbon atom, usually C-2, is ketonic.

K_i. See inhibition constant

kilobase pair (kb). A unit of length of double-stranded DNA, equivalent to 1000 base pairs.

kinase. An enzyme that catalyzes transfer of a phosphoryl group to an acceptor molecule. A protein kinase catalyzes the phosphorylation of protein substrates. Kinases are also known as phosphotransferases.

kinetic mechanism. A scheme used to describe the sequence of steps in a multisubstrate enzyme-catalyzed reaction.

kinetic order. The sum of the exponents in a rate equation, which reflects how many molecules are reacting in the slowest step of the reaction. Also known as reaction order.

kinetics. The study of rates of change, such as the rates of chemical reactions.

Klenow fragment. The C-terminal 605-residue fragment of *E. coli* DNA polymerase I produced by partial proteolysis. The Klenow fragment contains both the 5´→3´ polymerase and 3´→5´ proofreading exonuclease activities of DNA polymerase I but lacks the 5´→3´ exonuclease activity of the intact enzyme.

K_m. See Michaelis constant

Krebs cycle. See citric acid cycle

K_{tr}. See transport constant

K_w. See ion-product constant for water

lagging strand. The newly synthesized DNA strand formed by discontinuous 5´→3´ polymerization in the direction opposite replication fork movement.

lateral diffusion. The rapid motion of lipid or protein molecules within the plane of one leaflet of a lipid bilayer.

LDL. See low density lipoprotein

leader peptide. The peptide encoded by a portion of the leader region of an operon. Synthesis of a leader peptide is the basis for regulating transcription of the entire operon by the mechanism of attenuation.

leader region. The sequence of nucleotides that lie between the promoter and the first coding region of an operon.

leading strand. The newly synthesized DNA strand formed by continuous 5´→3´ polymerization in the same direction as replication fork movement.

leaflet. One layer of a lipid bilayer.

leaving group. The displaced group resulting from cleavage of a covalent bond.

lectin. A plant protein that binds specific saccharides in glycoproteins.

leucine zipper. A structural motif found in DNA-binding proteins and other proteins. The zipper is formed when the hydrophobic faces (frequently containing leucine residues) of two amphipathic α helices from the same or different polypeptide chains interact to form a coiled-coil structure.

LHC. See light-harvesting complex

ligand. A molecule, group, or ion that binds noncovalently to another molecule or atom.

ligand-gated ion channel. A membrane ion channel that opens or closes in response to binding of a specific ligand.

ligand-induced theory. See sequential theory of cooperativity and allosteric regulation

ligase. An enzyme that catalyzes the joining, or ligation, of two substrates. Ligation reactions require the input of the chemical potential energy of a nucleoside triphosphate such as ATP. Ligases are commonly referred to as synthetases.

light reactions. The photosynthetic reactions in which protons derived from water are used in the chemiosmotic synthesis of ATP from ADP + P_i and a hydride ion from water reduces $NADP^\oplus$ to NADPH. Also known as the light-dependent reactions.

light-dependent reactions. See light reactions

light-harvesting complex (LHC). A large pigment complex in the thylakoid membrane that aids a photosystem in gathering light.

light-independent reactions. See dark reactions

limit dextrin. A branched oligosaccharide derived from a glucose polysaccharide by the hydrolytic action of amylase or the phosphorolytic action of glycogen phosphorylase or starch phosphorylase. Limit dextrins are resistant to further degradation catalyzed by amylase or phosphorylase. Limit dextrins can be further degraded only after hydrolysis of the α-(1→6) linkages.

Lineweaver-Burk plot. See double-reciprocal plot

linker DNA. The stretch of DNA (approximately 54 base pairs) between two adjacent nucleosome core particles.

lipase. An enzyme that catalyzes the hydrolysis of triacylglycerols.

lipid. A water-insoluble (or sparingly soluble) organic compound found in biological systems, which can be extracted by using relatively nonpolar organic solvents.

lipid bilayer. A double layer of lipids in which the hydrophobic tails associate with one another in the interior of the bilayer and the polar head groups face outward into the aqueous environment. Lipid bilayers are the structural basis of biological membranes.

lipid vitamin. A polyprenyl compound composed primarily of a long hydrocarbon chain or fused ring. Unlike water-soluble vitamins, lipid vitamins can be stored by animals. Lipid vitamins include vitamins A, D, E, and K.

lipid-anchored membrane protein.   A membrane protein that is tethered to a membrane through covalent linkage to a lipid molecule.

lipolysis.   The metabolic hydrolysis of triacylglycerols.

lipopolysaccharide.   A macromolecule composed of lipid A (a disaccharide of phosphorylated glucosamine residues with attached fatty acids) and a polysaccharide. Lipopolysaccharides are found in the outer membrane of Gram-negative bacteria. These compounds are released from bacteria undergoing lysis and are toxic to humans and other animals. Also known as an endotoxin.

lipoprotein.   A macromolecular assembly of lipid and protein molecules with a hydrophobic core and a hydrophilic surface. Lipids are transported via lipoproteins.

liposome.   A synthetic vesicle composed of a phospholipid bilayer that encloses an aqueous compartment.

local regulators.   A class of short-lived eicosanoids that exert their regulatory effects near the cells in which they are produced.

loop.   A nonrepetitive polypeptide region that connects secondary structures within a protein molecule and provides directional changes necessary for a globular protein to attain its compact shape. Loops contain from 2 to 16 residues. Short loops of up to 5 residues are often called turns.

low density lipoprotein (LDL).   A type of plasma lipoprotein that is formed during the breakdown of IDLs and is enriched in cholesterol and cholesteryl esters.

low-barrier hydrogen bond.   A strong hydrogen bond in which the hydrogen is shared equally by two electronegative atoms that are less than 0.25 nm apart.

lumen.   The aqueous space enclosed by a biological membrane, such as the membrane of the endoplasmic reticulum or the thylakoid membrane.

lyase.   An enzyme that catalyzes a nonhydrolytic or nonoxidative elimination reaction, or lysis, of a substrate, with the generation of a double bond. In the reverse direction, a lyase catalyzes addition of one substrate to a double bond of a second substrate.

lysophosphoglyceride.   An amphipathic lipid that is produced when one of the two fatty acyl moieties of a glycerophospholipid is hydrolytically removed. Low concentrations of lysophosphoglycerides are metabolic intermediates, whereas high concentrations disrupt membranes, causing cells to lyse.

lysosome.   A specialized digestive organelle in eukaryotic cells. Lysosomes contain a variety of enzymes that catalyze the breakdown of cellular biopolymers, such as proteins, nucleic acids, and polysaccharides, and the digestion of large particles, such as some bacteria ingested by the cell.

major groove.   The wide groove on the surface of a DNA double helix created by the stacking of base pairs and the resulting twist in the sugar-phosphate backbones.

marker gene.   A gene, carried by a cloning vector, that can be used to distinguish between host cells that carry the vector and those that do not.

mass action ratio (Q).   The ratio of the concentrations of products to the concentrations of reactants of a reaction.

matrix.   See mitochondrial matrix

maximum velocity (V_{max}).   The initial velocity of a reaction when the enzyme is saturated with substrate, that is, when all the enzyme is in the form of an enzyme-substrate complex.

melting point (T_m).   The midpoint of the temperature range in which double-stranded DNA is converted to single-stranded DNA.

membrane.   A lipid bilayer containing associated proteins that serves to delineate and compartmentalize cells or organelles. Biological membranes are also the site of many important biochemical processes related to energy transduction and intracellular signaling.

membrane potential ($\Delta\psi$).   The charge separation across a membrane that results from differences in ionic concentrations on the two sides of the membrane.

messenger ribonucleic acid (mRNA).   A class of RNA molecules that serve as templates for protein synthesis.

metabolic fuel.   A small compound that can be catabolized to release energy. In multicellular organisms, metabolic fuels may be transported between tissues.

metabolically irreversible reaction.   A reaction in which the value of the mass action ratio is two or more orders of magnitude smaller than the value of the equilibrium constant. The free-energy change for such a reaction is a large negative number; thus, the reaction is irreversible.

metabolism.   The sum total of biochemical reactions carried out by an organism.

metabolite.   An intermediate in the synthesis or degradation of biopolymers and their component units.

metabolite channeling.   Transfer of the product of one reaction of a multifunctional enzyme or a multienzyme complex directly to the next active site or enzyme without entering the bulk solvent. Channeling increases the rate of a reaction pathway by decreasing the transit time for an intermediate to reach the next enzyme and by producing high local concentrations of the intermediate.

metabolite coenzyme.   A coenzyme synthesized from a common metabolite.

metal-activated enzyme.   An enzyme that either has an absolute requirement for metal ions or is stimulated by the addition of metal ions.

metalloenzyme.   An enzyme that contains one or more firmly bound metal ions. In some cases, such metal ions constitute part of the active site of the enzyme and are active participants in catalysis.

micelle.   An aggregation of amphipathic molecules in which the hydrophilic portions of the molecules project into the aqueous environment and the hydrophobic portions associate with one another in the interior of the structure to minimize contact with water molecules.

Michaelis constant (K_m).   The concentration of substrate that results in an initial velocity (v_0) equal to one-half the maximum velocity (V_{max}) for a given reaction.

Michaelis-Menten equation.   A rate equation relating the initial velocity (v_0) of an enzymatic reaction to the substrate concentration ([S]), the maximum velocity (V_{max}), and the Michaelis constant (K_m).

microfilament.   See actin filament

microtubule.   A protein filament composed of a and b tubulin heterodimers. Microtubules are components of the cytoskeletal network and can form structures capable of directed movement.

minor groove.   The narrow groove on the surface of a DNA double helix created by the stacking of base pairs and the resulting twist in the sugar-phosphate backbones.

mismatch repair. Restoration of the normal nucleotide sequence in a DNA molecule containing mismatched bases. In mismatch repair, the correct strand is recognized, a portion of the incorrect strand is excised, and correctly base-paired, double-stranded DNA is synthesized by the actions of DNA polymerase and DNA ligase.

missense mutation. An alteration in DNA that involves the substitution of one nucleotide for another, resulting in a change in the amino acid specified by that codon.

mitochondrial matrix. The gel-like phase enclosed by the inner membrane of the mitochondrion. The mitochondrial matrix contains many enzymes involved in aerobic energy metabolism.

mitochondrion. An organelle that is the main site of oxidative energy metabolism in most eukaryotic cells. Mitochondria contain an outer and an inner membrane, the latter characteristically folded into cristae.

mixed micelle. A micelle containing more than one type of amphipathic molecule.

mixed oligonucleotide probe. Short stretches of DNA corresponding to each of the possible sequences that could encode a given protein sequence. A mixed oligonucleotide probe is used to screen a DNA library for the presence of the protein-encoding clone.

molar mass. The weight in grams of one mole of a compound.

molecular weight. See relative molecular mass

molecular-exclusion chromatography. See gel-filtration chromatography

monocistronic mRNA. An mRNA molecule that encodes only a single polypeptide. Most eukaryotic mRNA molecules are monocistronic.

monomer. 1. A small compound that becomes a residue when polymerized with other monomers. 2. A single subunit of a multisubunit protein.

monosaccharide. A simple sugar of three or more carbon atoms with the empirical formula $(CH_2O)_n$.

monotopic protein. An integral membrane protein anchored by a single membrane-spanning segment.

motif. A combination of secondary structure that appears in a number of different proteins. Also known as supersecondary structure.

M_r. See relative molecular mass

mRNA. See messenger ribonucleic acid

mRNA precursor. A class of RNA molecules synthesized by eukaryotic RNA polymerase II. mRNA precursors are processed posttranscriptionally to produce messenger RNA. Formerly known as heterogeneous nuclear RNA (hnRNA).

mucin. A high-molecular-weight O-linked glycoprotein containing as much as 80% carbohydrate by mass. Mucins are extended, negatively charged molecules that contribute to the viscosity of mucus, the fluid found on the surfaces of the gastrointestinal, genitourinary, and respiratory tracts.

multienzyme complex. An oligomeric protein that catalyzes several metabolic reactions.

mutagen. An agent that can cause DNA damage.

mutation. A heritable change in the sequence of nucleotides in DNA that causes a permanent alteration of genetic information.

near-equilibrium reaction. A reaction in which the value of the mass action ratio is close to the value of the equilibrium constant. The free-energy change for such a reaction is small;

thus, the reaction is reversible.

Nernst equation. An equation that relates the observed change in reduction potential (ΔE) to the change in standard reduction potential ($\Delta E°'$) of a reaction.

neutral solution. An aqueous solution that has a pH value of 7.0.

nick translation. The process in which DNA polymerase binds to a gap between the 3′ end of a nascent DNA chain and the 5′ end of the next RNA primer, catalyzes hydrolytic removal of ribonucleotides using $5′\rightarrow 3′$ exonuclease activity, and replaces them with deoxyribonucleotides using $5′\rightarrow 3′$ polymerase activity.

nitrogen cycle. The flow of nitrogen from N_2 to nitrogen oxides ($NO_2^{\ominus}$ and $NO_3^{\ominus}$), ammonia, nitrogenous biomolecules, and back to N_2.

nitrogen fixation. The reduction of atmospheric nitrogen to ammonia. Biological nitrogen fixation occurs in only a few species of bacteria and algae.

N-linked glycoprotein. A glycoprotein in which one or more oligosaccharide chains are attached to the protein through covalent bonds to the amide nitrogen atom of the side chain of asparagine residues. The oligosaccharide chains of N-linked glycoproteins contain a core pentasaccharide of two N-acetylglucosamine residues and three mannose residues.

NMR spectroscopy. See nuclear magnetic resonance spectroscopy

noncompetitive inhibition. Inhibition of an enzyme-catalyzed reaction by an inhibitor that binds to either the enzyme or the enzyme-substrate complex.

nonessential amino acid. An amino acid that an animal can produce in sufficient quantity to meet metabolic needs.

nonrepetitive structure. An element of protein structure in which consecutive residues do not have a single repeating conformation. Also known as a random coil.

nonsense mutation. An alteration in DNA that involves the substitution of one nucleotide for another, changing a codon that specifies an amino acid to a termination codon. A nonsense mutation results in premature termination of a protein's synthesis.

N-terminus. The amino acid residue bearing a free a-amino group at one end of a peptide chain. In some proteins, the N-terminus is blocked by acylation. The N-terminal residue is usually assigned the residue number 1. Also known as the amino terminus.

nuclear envelope. The double membrane that surrounds the nucleus and contains protein-lined nuclear pore complexes that regulate the import and export of material to and from the nucleus. The outer membrane of the nuclear envelope is continuous with the endoplasmic reticulum; the inner membrane is lined with filamentous proteins, constituting the nuclear lamina.

nuclear magnetic resonance spectroscopy (NMR spectroscopy). A technique used to study the structures of molecules in solution. In nuclear magnetic resonance spectroscopy, the absorption of electromagnetic radiation by molecules in magnetic fields of varying frequencies is used to determine the spin states of certain atomic nuclei.

nuclease. An enzyme that catalyzes hydrolysis of the phosphodiester linkages of a polynucleotide chain. Nucleases can be classified as endonucleases and exonucleases.

nuclease protection. See footprinting

nucleic acid. A polymer composed of nucleotide residues linked in a linear sequence by 3′-5′ phosphodiester bonds. DNA and RNA are nucleic acids composed of deoxyribonucleotide residues and ribonucleotide residues, respectively.

nucleoid region. The region within a prokaryotic cell that contains the chromosome.

nucleolus. The region of the eukaryotic nucleus where rRNA transcripts are processed and ribosomes are assembled.

nucleophile. An electron-rich species that is negatively charged or contains unshared electron pairs and is attracted to chemical species that are positively charged or electron-deficient (electrophiles).

nucleoside. A purine or pyrimidine N-glycoside of ribose or deoxyribose.

nucleosome. A DNA-protein complex that forms the fundamental unit of chromatin. A nucleosome consists of a nucleosome core particle (approximately 146 base pairs of DNA plus a histone octamer), linker DNA (approximately 54 base pairs), and histone H1 (which binds the core particle and linker DNA).

nucleosome core particle. A DNA-protein complex composed of approximately 146 base pairs of DNA wrapped around an octamer of histones (two each of H2A, H2B, H3, and H4).

nucleotide. The phosphate ester of a nucleoside, consisting of a nitrogenous base linked to a pentose phosphate. Nucleotides are the monomeric units of nucleic acids.

nucleotide probe. A labeled oligonucleotide used to screen DNA or RNA molecules for the presence of a specific complementary sequence.

nucleus. An organelle that contains the principal genetic material of eukaryotic cells and functions as the major site of RNA synthesis and processing.

obligate aerobe. An organism that requires the presence of oxygen for survival.

obligate anaerobe. An organism that requires an oxygen-free environment for survival.

obligatory glycolytic tissue. A tissue with little or no capacity for oxidative metabolism and for which glucose is the only usable fuel.

Okazaki fragments. Relatively short strands of DNA that are produced during discontinuous synthesis of the lagging strand of DNA.

oligomer. A multisubunit molecule whose arrangement of subunits always has a defined stoichiometry and almost always displays symmetry.

oligonucleotide. A polymer of several (up to about 20) nucleotide residues linked by phosphodiester bonds.

oligopeptide. A polymer of several (up to about 20) amino acid residues linked by peptide bonds.

oligosaccharide. A polymer of 2 to about 20 monosaccharide residues linked by glycosidic bonds.

oligosaccharide processing. The enzyme-catalyzed addition and removal of saccharide residues during the maturation of a glycoprotein.

***O*-linked glycoprotein.** A glycoprotein in which one or more oligosaccharide chains are attached to the protein through covalent bonds, usually to the hydroxyl oxygen atom of serine or threonine residues.

oncogene. A gene whose product has the ability to transform normal eukaryotic cells into cancer cells. Some oncogenes are carried by viruses.

open reading frame. A stretch of nucleotide triplets that contains no termination codons. Protein-encoding regions are examples of open reading frames.

operator. A DNA sequence to which a specific repressor protein binds, thereby blocking transcription of a gene or operon.

operon. A bacterial transcriptional unit consisting of several different genes cotranscribed from one promoter.

ordered reaction. A reaction in which both the binding of substrates to an enzyme and the release of products from the enzyme follow an obligatory order.

organ. An association of one or more tissues that carry out discrete functions within a multicellular organism.

organelle. Any specialized membrane-bounded structure within a eukaryotic cell. Organelles are uniquely organized to perform specific functions.

origin of replication. A DNA sequence at which replication is initiated.

osmosis. The movement of solvent molecules from a less concentrated solution to an adjacent, more concentrated solution.

osmotic pressure. The pressure required to prevent the flow of solvent from a less concentrated solution to a more concentrated solution.

oxidase. An enzyme that catalyzes an oxidation-reduction reaction in which O_2 is the electron acceptor. Oxidases are members of the IUB class of enzymes known as oxidoreductases.

oxidation. The loss of electrons from a substance through transfer to another substance (the oxidizing agent). Oxidations can take several forms, including the addition of oxygen to a compound, the removal of hydrogen from a compound to create a double bond, or an increase in the valence of a metal ion.

oxidative phosphorylation. A set of reactions in which compounds such as NADH and reduced ubiquinone (QH_2) are aerobically oxidized and ATP is generated from ADP and P_i. Oxidative phosphorylation consists of two tightly coupled phenomena: oxidation of substrates by the respiratory electron-transport chain, accompanied by the translocation of protons across the inner mitochondrial membrane to generate a proton concentration gradient; and formation of ATP, driven by the flux of protons into the matrix through a channel in $F_O F_1$ ATP synthase.

oxidizing agent. A substance that accepts electrons in an oxidation-reduction reaction and thereby becomes reduced.

oxidoreductase. An enzyme that catalyzes an oxidation-reduction reaction. Some oxidoreductases are known as dehydrogenases, oxidases, peroxidases, oxygenases, or reductases.

oxygenase. An enzyme that catalyzes incorporation of molecular oxygen into a substrate. Oxygenases are members of the IUB class of enzymes known as oxidoreductases.

oxygenation. The reversible binding of oxygen to a macromolecule.

P site. See peptidyl site

Δp. See protonmotive force

PAGE. See polyacrylamide gel electrophoresis

pancreatic juice. A solution of sodium chloride, zymogens, and digestive enzymes secreted by the pancreas during digestion. Pancreatic juice may also contain sodium bicarbonate.

passive transport. The process by which a solute specifically binds to a transport protein and is transported across a membrane, moving with the solute concentration gradient. Passive transport occurs without the expenditure of energy. Also known as facilitated diffusion.

Pasteur effect. The slowing of glycolysis in the presence of oxygen.

pathway. A sequence of metabolic reactions.

pause site. A region of a gene where transcription slows. Pausing is exaggerated at palindromic sequences, where newly synthesized RNA can form a hairpin structure.

PCR. See polymerase chain reaction

pentose phosphate pathway. A pathway by which glucose 6-phosphate is metabolized to generate NADPH and ribose 5-phosphate. In the oxidative stage of the pathway, glucose 6-phosphate is converted to ribulose 5-phosphate and CO_2, generating two molecules of NADPH. In the nonoxidative stage, ribulose 5-phosphate can be isomerized to ribose 5-phosphate or converted to intermediates of glycolysis. Also known as the hexose monophosphate shunt.

peptide. Two or more amino acids covalently joined in a linear sequence by peptide bonds.

peptide bond. The covalent secondary amide linkage that joins the carbonyl group of one amino acid residue to the amino nitrogen of another in peptides and proteins.

peptide group. The nitrogen and carbon atoms involved in a peptide bond and their four substituents: the carbonyl oxygen atom, the amide hydrogen atom, and the two adjacent a-carbon atoms.

peptidoglycan. A macromolecule containing a heteroglycan chain of alternating *N*-acetylglucosamine and *N*-acetylmuramic acid cross-linked to peptides of varied composition. Peptidoglycans are the major components of the cell walls of many bacteria.

peptidyl site. The site on a ribosome that is occupied during protein synthesis by the tRNA molecule attached to the growing polypeptide chain. Also known as the P site.

peptidyl transferase. The enzymatic activity responsible for the formation of a peptide bond during protein synthesis.

peptidyl-tRNA. The tRNA molecule to which the growing peptide chain is attached during protein synthesis.

peripheral membrane protein. A membrane protein that is weakly bound to the interior or exterior surface of a membrane through ionic interactions and hydrogen bonding with the polar heads of the membrane lipids or with an integral membrane protein. Also known as an extrinsic membrane protein.

periplasmic space. The region between the plasma membrane and the cell wall in bacteria.

permeability coefficient. A measure of the ability of an ion or small molecule to diffuse across a lipid bilayer.

peroxidase. An enzyme that catalyzes a reaction in which hydrogen peroxide (H_2O_2) is the oxidizing agent. Peroxidases are members of the IUB class of enzymes known as oxidoreductases.

peroxisome. An organelle in all animal and many plant cells that carries out oxidation reactions, some of which produce the toxic compound hydrogen peroxide (H_2O_2). Peroxisomes contain the enzyme catalase, which catalyzes the breakdown of toxic H_2O_2 to water and O_2.

pH. A logarithmic quantity that indicates the acidity of a solution, that is, the concentration of hydronium ions in solution. pH is defined as the negative logarithm of the hydronium ion concentration.

pH optimum. In an enzyme-catalyzed reaction, the pH at the point of maximum catalytic activity.

phage. See bacteriophage

phase-transition temperature (T_m). The midpoint of the temperature range in which lipids or other macromolecular aggregates are converted from a highly ordered phase or state (such as a gel) to a less-ordered state (such as a liquid crystal).

ϕ (phi). The angle of rotation around the bond between the α-carbon and the nitrogen of a peptide group.

phosphagen. A high-energy phosphate-storage molecule found in animal muscle cells. Phosphagens are phosphoamides and have a higher phosphoryl-group–transfer potential than ATP.

phosphatase. An enzyme that catalyzes the hydrolytic removal of a phosphoryl group from a phosphoprotein.

phosphatidate. A glycerophospholipid that consists of two fatty acyl groups esterified to C-1 and C-2 of glycerol 3-phosphate. Phosphatidates are metabolic intermediates in the biosynthesis or breakdown of more complex glycerophospholipids.

phosphatidylinositol-glycan–linked glycoprotein. A glycoprotein in which the protein is attached to a phosphoethanolamine moiety that is linked to a branched oligosaccharide to which the lipid phosphatidylinositol is also attached. The phosphatidylinositol-glycan structure is known as a GPI membrane anchor.

phosphoanhydride. A compound formed by condensation of two phosphate groups.

phosphodiester linkage. A linkage in nucleic acids and other molecules in which two alcoholic hydroxyl groups are joined through a phosphate group.

phosphoester linkage. The bond by which a phosphoryl group is attached to an alcoholic or phenolic oxygen.

phospholipid. A lipid containing a phosphate moiety.

phosphorolysis. Cleavage of a bond within a molecule by group transfer to an oxygen atom of phosphate.

phosphorylase. An enzyme that catalyzes the cleavage of its substrate(s) via nucleophilic attack by inorganic phosphate (P_i) (i.e., via phosphorolysis).

phosphorylation. A reaction involving the addition of a phosphoryl group to a molecule.

phosphoryl-group–transfer potential. A measure of the ability of a compound to transfer a phosphoryl group to another compound. Under standard conditions, group-transfer potentials have the same values as the standard free energies of hydrolysis but are opposite in sign.

phosphotransferase. See kinase

photon. A quantum of light energy.

photophosphorylation. The light-dependent formation of ATP from ADP and P_i catalyzed by chloroplast ATP synthase.

photoreactivation. The direct repair of damaged DNA by an enzyme that is activated by visible light.

photorespiration. The light-dependent uptake of O_2 and the subsequent metabolism of phosphoglycolate that occurs primarily in C_3 photosynthetic plants. Photorespiration can occur because O_2 competes with CO_2 for the active site of ribulose 1,5-bisphosphate carboxylase-oxygenase, the enzyme that catalyzes the first step of the reductive pentose phosphate cycle.

photosynthesis.   The process by which carbohydrates are synthesized from atmospheric CO_2 and water using light as the source of energy.

photosynthetic carbon reduction cycle.   See reductive pentose phosphate cycle

photosystem.   A functional unit of the light-dependent reactions of photosynthesis. Each membrane-embedded photosystem contains a reaction center, which forms the core of the photosystem, and a pool of light-absorbing antenna pigments.

phototroph.   An organism that can convert light energy into chemical potential energy (i.e., an organism capable of photosynthesis).

physiological pH.   The normal pH of human blood, which is 7.4.

pI.   See isoelectric point

ping-pong reaction.   A reaction in which an enzyme binds one substrate and releases a product, leaving a substituted enzyme that then binds a second substrate and releases a second product, thereby restoring the enzyme to its original form.

pitch.   The axial distance for one complete turn of a helical structure.

pK_a.   A logarithmic value that indicates the strength of an acid. pK_a is defined as the negative logarithm of the acid dissociation constant, K_a.

plaque.   An area of dead or slowly growing bacterial cells that indicates the presence of bacteriophage-infected cells among uninfected cells.

plasma membrane.   The membrane that surrounds the cytoplasm of a cell and thus defines the perimeter of the cell.

plasmalogen.   A glycerophospholipid that has a hydrocarbon chain linked to C-1 of glycerol 3-phosphate through a vinyl ether linkage. Plasmalogens are found in the central nervous system and in peripheral nerve and muscle tissue.

plasmid.   A relatively small, extrachromosomal DNA molecule in bacteria and yeast that is capable of autonomous replication. Plasmids are closed, circular, double-stranded DNA molecules.

P:O ratio.   The ratio of molecules of ADP phosphorylated to atoms of oxygen reduced during oxidative phosphorylation.

polar.   Having uneven distribution of charge. A molecule or functional group is polar if its center of negative charge does not coincide with its center of positive charge.

poly A tail.   A stretch of polyadenylate, up to 250 nucleotide residues long, that is added to the 3′ end of a eukaryotic mRNA molecule following transcription.

polyacrylamide gel electrophoresis (PAGE).   A technique used to separate molecules of different net charge and/or size based on their migration through a highly cross-linked gel matrix in an electric field.

polycistronic mRNA.   An mRNA molecule that contains multiple coding regions. Many prokaryotic mRNA molecules are polycistronic.

polymerase chain reaction (PCR).   A method for amplifying the amount of DNA in a sample and for enriching a particular DNA sequence in a population of DNA molecules. In the polymerase chain reaction, oligonucleotides complementary to the ends of the desired DNA sequence are used as primers for multiple rounds of DNA synthesis.

polynucleotide.   A polymer of many (usually more than 20) nucleotide residues linked by phosphodiester bonds.

polypeptide.   A polymer of many (usually more than 20) amino acid residues linked by peptide bonds.

polyribosome.   See polysome

polysaccharide.   A polymer of many (usually more than 20) monosaccharide residues linked by glycosidic bonds. Polysaccharide chains can be linear or branched.

polysome.   The structure formed by the binding of many translation complexes to a large mRNA molecule. Also known as a polyribosome.

polytene chromosome.   A chromosome that is replicated many times without separation of the copies. The resulting structure is composed of regions of condensed chromatin and regions expanded to form puffs.

polytopic protein.   An integral membrane protein anchored by multiple membrane-spanning segments.

pore.   See channel

postabsorptive phase.   The period following the absorptive phase. During the postabsorptive phase, which lasts about 12 hours in humans, glucose is mobilized from glycogen stores and is synthesized via gluconeogenesis.

posttranscriptional processing.   RNA processing that occurs after transcription is complete.

posttranslational modification.   Covalent modification of a protein that occurs after elongation of the polypeptide is complete.

prenylated protein.   A lipid-anchored protein that is covalently linked to an isoprenoid moiety via the sulfur atom of a cysteine residue at the C-terminus of the protein.

primary structure.   The sequence in which residues are covalently linked to form a polymeric chain.

primary transcript.   A newly synthesized RNA molecule before processing.

primase.   An enzyme in the primosome that catalyzes the synthesis of short pieces of RNA about 10 residues long. These oligonucleotides are the primers for synthesis of Okazaki fragments.

primosome.   A multiprotein complex, including primase and helicase in *E. coli*, that catalyzes the synthesis of the short RNA primers needed for discontinuous DNA synthesis of the lagging strand.

probe.   See nucleotide probe

processive enzyme.   An enzyme that remains bound to its growing polymeric product through many polymerization steps.

prochiral atom.   An atom with multiple substituents, two of which are identical. A prochiral atom can become chiral when one of the identical substituents is replaced.

proenzyme.   See zymogen

prokaryote.   An organism, usually a single cell, which contains no nucleus or internal membranes.

promoter.   The region of DNA where RNA polymerase binds during transcription initiation.

prostaglandin.   An eicosanoid that has a cyclopentane ring. Prostaglandins are metabolic regulators that act in the immediate neighborhood of the cells in which they are produced.

prosthetic group.   A coenzyme that is tightly bound to an enzyme. A prosthetic group, unlike a cosubstrate, remains bound to a specific site of the enzyme throughout the catalytic cycle of the enzyme.

protease.   An enzyme that catalyzes hydrolysis of peptide bonds. The physiological substrates of proteases are proteins.

protein.   A biopolymer consisting of one or more polypeptide chains. The biological function of each protein molecule depends not only on the sequence of covalently linked amino acid residues, but also on its three-dimensional structure (conformation).

protein coenzyme.   A protein that does not itself catalyze reactions but is required for the action of certain enzymes. Also known as a group-transfer protein.

protein glycosylation.   The covalent addition of carbohydrate to proteins. In *N*-glycosylation, the carbohydrate is attached to the amide group of the side chain of an asparagine residue. In *O*-glycosylation, the carbohydrate is attached to the hydroxyl group of the side chain of a serine or threonine residue.

protein kinase.   See kinase

protein phosphatase.   See phosphatase

proteoglycan.   A complex of protein with glycosaminoglycan chains covalently bound through their anomeric carbon atoms. Up to 95% of the mass of a proteoglycan may be glycosaminoglycan.

protonmotive force (Δp).   The energy stored in a proton concentration gradient across a membrane.

proto-oncogene.   A eukaryotic gene that can be mutated to an oncogene.

proximity effect.   The increase in the rate of a nonenzymatic or enzymatic reaction attributable to high effective concentrations of reactants, which result in more frequent formation of transition states.

pseudo first-order reaction.   A multi-reactant reaction carried out under conditions where the rate depends on the concentration of only one reactant.

pseudogene.   A nonexpressed sequence of DNA that evolved from a protein-encoding gene. Pseudogenes often contain mutations in their coding regions and cannot produce functional proteins.

ψ (psi).   The angle of rotation around the bond between the α-carbon and the carbonyl carbon of a peptide group.

Δψ.   See membrane potential

purine.   A nitrogenous base having a two-ring structure in which a pyrimidine is fused to imidazole. Adenine and guanine are substituted purines found in both DNA and RNA.

purine nucleotide cycle.   A pathway in muscle that converts aspartate to fumarate and ammonia.

pyranose.   A monosaccharide structure that forms a six-membered ring as a result of intramolecular hemiacetal formation.

pyrimidine.   A nitrogenous base having a heterocyclic ring that consists of four carbon atoms and two nitrogen atoms. Cytosine, thymine, and uracil are substituted pyrimidines found in nucleic acids (cytosine in DNA and RNA, uracil in RNA, and thymine principally in DNA).

Q.   See mass action ratio

Q cycle.   A cyclic pathway proposed to explain the sequence of electron transfers and proton movements within Complex III of mitochondria or the cytochrome bf complex in chloroplasts. The net result of the two steps of the Q cycle is oxidation of two molecules of QH_2 or plastoquinol(PQH_2); formation of one molecule of QH_2 or PQH_2; transfer of two electrons; and net translocation of four protons across the inner mitochondrial membrane to the intermembrane space or across the thylakoid membrane to the lumen.

quaternary structure.   The organization of two or more polypeptide chains within a multisubunit protein.

R group.   A part of a molecule not explicitly shown in a chemical structure, such as an alkyl group or the side chain of an amino acid.

R state.   The more active conformation of an allosteric protein; opposite of T state.

Ramachandran plot.   A plot of ψ versus φ values for amino acid residues in a polypeptide chain. Certain φ and ψ values are characteristic of different conformations.

random coil.   See nonrepetitive structure

random reaction.   A reaction in which neither the binding of substrates to an enzyme nor the release of products from the enzyme follows an obligatory order.

rate acceleration.   The ratio of the rate constant for a reaction in the presence of enzyme (k_{cat}) divided by the rate constant for that reaction in the absence of enzyme (k_n). The rate acceleration value is a measure of the efficiency of an enzyme.

rate equation.   An expression of the observed relationship between the velocity of a reaction and the concentration of each reactant.

rate-determining step.   The slowest step in a chemical reaction. The rate-determining step has the highest activation energy among the steps leading to formation of a product from the substrate.

reaction center.   A complex of proteins, electron-transporting cofactors, and a special pair of chlorophyll molecules that forms the core of a photosystem. The reaction center is the site of conversion of photochemical energy to electrochemical energy during photosynthesis.

reaction mechanism.   The step-by-step atomic or molecular events that occur during chemical reactions.

reaction order.   See kinetic order

reaction specificity.   The lack of formation of wasteful byproducts by an enzyme. Reaction specificity results in essentially 100% product yields.

reactive center.   The part of a coenzyme to which mobile metabolic groups are attached.

reading frame.   The sequence of nonoverlapping codons of an mRNA molecule that specifies the amino acid sequence. The reading frame of an mRNA molecule is determined by the position where translation begins; usually an AUG codon.

receptor.   A cell-surface protein that binds a specific ligand, leading to some cellular response.

recombinant DNA.   A DNA molecule that includes DNA from different sources.

recombinant DNA technology.   The methodologies for isolating, manipulating, and amplifying identifiable sequences of DNA. Also known as genetic engineering.

recombination.   See genetic recombination

reducing agent.   A substance that loses electrons in an oxidation-reduction reaction and thereby becomes oxidized.

reducing end.   The residue containing a free anomeric carbon in a polysaccharide. A polysaccharide usually contains no more than one reducing end.

reductase.   See oxidoreductase

reduction.   The gain of electrons by a substance through transfer from another substance (the reducing agent). Reductions can take several forms, including the loss of oxygen from a compound, the addition of hydrogen to a double bond of a compound, or a decrease in the valence of a metal ion.

reduction potential (E).   A measure of the tendency of a substance to reduce other substances. The more negative the reduction potential, the greater the tendency to donate electrons.

reductive pentose phosphate cycle (RPP cycle). A cycle of reactions that leads to the reductive conversion of carbon dioxide to carbohydrates during photosynthesis. Also known as the C_3 pathway, the photosynthetic carbon reduction cycle, and the Calvin-Benson cycle.

regulatory enzyme. An enzyme located at a critical point within one or more metabolic pathways, whose activity may be increased or decreased based on metabolic demand. Most regulatory enzymes are oligomeric.

regulatory site. A ligand-binding site in a regulatory enzyme distinct from the active site. Allosteric modulators alter enzyme activity by binding to the regulatory site. Also known as an allosteric site.

regulon. A group of coordinately regulated operons or genes.

relative molecular mass (M_r). The mass of a molecule relative to $^1/_{12}$th the mass of ^{12}C.

release factor. A protein involved in terminating protein synthesis.

renaturation. The restoration of the native conformation of a biological macromolecule, usually resulting in restoration of biological activity.

replication. The duplication of double-stranded DNA, during which parental strands separate and serve as templates for synthesis of new strands. Replication is carried out by DNA polymerase and associated factors.

replication fork. The Y-shaped junction where double-stranded, template DNA is unwound and new DNA strands are synthesized during replication.

replisome. A multiprotein complex that includes DNA polymerase, primase, helicase, single-strand binding protein, and additional components. The replisomes, located at each of the replication forks, carry out the polymerization reactions of bacterial chromosomal DNA replication.

repressor. A regulatory DNA-binding protein that prevents transcription by RNA polymerase.

residue. A single component within a polymer. The chemical formula of a residue is that of the corresponding monomer minus the elements of water.

respiratory electron-transport chain. A series of enzyme complexes and associated cofactors that are electron carriers, passing electrons from reduced coenzymes or substrates to molecular oxygen (O_2), the terminal electron acceptor of aerobic metabolism.

respiratory-control index. The ratio in mitochondria of the rate of oxygen consumption in the presence of ADP to the rate in the absence of ADP.

restriction endonuclease. An endonuclease that catalyzes the hydrolysis of double-stranded DNA at a specific nucleotide sequence. Type I restriction endonucleases catalyze both the methylation of host DNA and the cleavage of nonmethylated DNA, whereas type II restriction endonucleases catalyze only the cleavage of nonmethylated DNA.

restriction fragment length polymorphism (RFLP). Variation in the length of DNA fragments generated by the action of restriction endonucleases on genomic DNA from different individuals.

restriction map. A diagram showing the size and arrangement of fragments produced from a DNA molecule by the action of various restriction endonucleases.

retrovirus. An RNA virus that infects a eukaryotic cell.

reverse transcriptase. A type of DNA polymerase that catalyzes the synthesis of a strand of DNA from an RNA template.

RFLP. See restriction fragment length polymorphism

ribonuclease (RNase). An enzyme that catalyzes the hydrolysis of ribonucleic acids to form oligonucleotides and/or mononucleotides.

ribonucleic acid (RNA). A polymer consisting of ribonucleotide residues joined by 3′–5′ phosphodiester bonds. The sugar moiety in RNA is ribose. Genetic information contained in DNA is transcribed in the synthesis of RNA, some of which (mRNA) is translated in the synthesis of protein.

ribonucleoprotein. A complex containing both ribonucleic acid and protein.

ribophorin. A ribosome-binding protein that anchors the ribosome to the endoplasmic reticulum surface and through which proteins are translocated into the lumen as they are synthesized.

ribose. A five-carbon monosaccharide ($C_5H_{10}O_5$) that is the carbohydrate component of RNA, ATP, and numerous coenzymes. 2-Deoxyribose is the carbohydrate component of DNA.

ribosomal ribonucleic acid (rRNA). A class of RNA molecules that are integral components of ribosomes. rRNA is the most abundant cellular RNA.

ribosome. A large ribonucleoprotein complex composed of multiple rRNA molecules and proteins. Ribosomes are the site of protein synthesis.

ribozyme. An RNA molecule with enzymatic activity.

rise. The distance between one residue and the next along the axis of a helical macromolecule.

R-looping. A technique used to visualize the organization of a gene. In R-looping, mature RNA is hybridized to DNA corresponding to the gene, and stretches of DNA that are not represented in the RNA sequence, such as introns, appear as single-stranded loops in an electron micrograph.

RNA. See ribonucleic acid

RNA editing. Posttranscriptional changes in the coding region of an mRNA molecule. Editing may include nucleotide modifications and insertions or deletions.

RNA processing. The reactions that transform a primary RNA transcript into a mature RNA molecule. The three general types of RNA processing include the removal of RNA nucleotides from primary transcripts, the addition of RNA nucleotides not encoded by the gene, and the covalent modification of bases.

RNase. See ribonuclease

RPP cycle. See reductive pentose phosphate cycle

rRNA. See ribosomal ribonucleic acid

S. See Svedberg unit

S. See entropy

saccharide. See carbohydrate

salvage pathway. A pathway in which a major metabolite, such as a purine or pyrimidine nucleotide, can be synthesized from a preformed molecular entity, such as a purine or pyrimidine.

saturated fatty acid. A fatty acid that does not contain a carbon-carbon double bond.

Schiff base. A complex formed by the reversible condensation of a primary amine with an aldehyde (to form an aldimine) or a ketone (to form a ketimine).

screen. A test of transformed cells for the presence of a desired recombinant DNA molecule.

SDS-PAGE. See sodium dodecyl sulfate–polyacrylamide gel electrophoresis

second messenger. A compound that acts intracellularly in response to an extracellular signal.

secondary structure. The regularities in local conformations within macromolecules. In proteins, secondary structure is maintained by hydrogen bonds between carbonyl and amide groups of the backbone. In nucleic acids, secondary structure is maintained by hydrogen bonds and stacking interactions between the bases.

second-order reaction. A reaction whose rate depends on the concentrations of two reactants.

secretory vesicle. A vesicle carrying proteins destined for secretion. Secretory vesicles bud off the Golgi apparatus and travel to the plasma membrane where the vesicle contents are released by exocytosis.

selectin. A membrane glycoprotein containing a lectinlike domain that binds oligosaccharide groups. CaH is required for selectin binding to oligosaccharides.

selection. A technique in which only transformed cells survive under a chosen set of experimental conditions.

self-splicing intron. An intron that is excised in a reaction mediated by the RNA precursor itself.

semiconservative replication. The mode of duplicating DNA in which each strand serves as a template for the synthesis of a complementary strand. The result is two molecules of double-stranded DNA, each of which contains one of the parental strands.

sequential reaction. An enzymatic reaction in which all the substrates must be bound to the enzyme before any product is released.

sequential theory of cooperativity and allosteric regulation. A model of the cooperative binding of identical ligands to oligomeric proteins. According to the simplest form of the sequential theory, the binding of a ligand may induce a change in the tertiary structure of the subunit to which it binds and may alter the conformations of neighboring subunits to varying extents. Only one subunit conformation has a high affinity for the ligand. Also known as the ligand-induced theory.

serine protease. A protease with an active-site serine residue that acts as a nucleophile during catalysis.

Shine-Dalgarno sequence. A purine-rich region just upstream of the initiation codon in prokaryotic mRNA molecules. The Shine-Dalgarno sequence binds to a pyrimidine-rich sequence in the ribosomal RNA, thereby positioning the ribosome at the initiation codon.

shuttle vector. A cloning vector that can replicate in both prokaryotic and eukaryotic cells. Shuttle vectors are used to transfer recombinant DNA molecules between prokaryotic and eukaryotic cells.

σ cascade. The sequential expression of different s subunits that bind to the core RNA polymerase. A σ cascade can regulate gene expression since different σ subunits can direct RNA polymerase to different promoters, including those for genes encoding additional σ subunits.

σ factor. See σ subunit

σ subunit. A subunit of prokaryotic RNA polymerase, which acts as a transcription initiation factor by binding to the promoter. Different σ subunits are specific for different promoters. Also known as a σ factor.

signal peptidase. An integral membrane protein of the endoplasmic reticulum that catalyzes cleavage of the signal peptide of proteins translocated to the lumen.

signal peptide. The N-terminal sequence of residues in a newly synthesized polypeptide that targets the protein for translocation across a membrane.

signal transduction. The process whereby an extracellular signal is converted to an intracellular signal by the action of a membrane-associated receptor, a transducer, and an effector enzyme.

signal-recognition particle (SRP). A eukaryotic protein-RNA complex that binds a newly synthesized peptide as it is extruded from the ribosome. The signal-recognition particle is involved in anchoring the ribosome to the cytosolic face of the endoplasmic reticulum so that protein translocation to the lumen can occur.

single-strand binding protein (SSB). A protein that binds tightly to single-stranded DNA, preventing the DNA from folding back on itself to form double-stranded regions. Also known as helix-destabilizing protein.

site-directed mutagenesis. An in vitro procedure by which one particular nucleotide residue in a gene is replaced by another, resulting in production of an altered protein sequence.

small nuclear ribonucleoprotein (snRNP). An RNA-protein complex composed of one or two specific snRNA molecules plus a number of proteins. snRNPs are involved in splicing mRNA precursors and in other cellular events.

small RNA. A class of RNA molecules that participate in RNA processing. Some small RNA molecules have catalytic activity. Some small nuclear RNA molecules (snRNA) are components of small nuclear ribonucleoproteins (snRNPs).

snRNA. See small nuclear RNA

snRNP. See small nuclear ribonucleoprotein

soap. An alkali metal salt of a long-chain fatty acid. Soaps are a type of detergent.

sodium dodecyl sulfate–polyacrylamide gel electrophoresis (SDS-PAGE). Polyacrylamide gel electrophoresis performed in the presence of the detergent sodium dodecyl sulfate. SDS-PAGE allows separation of proteins on the basis of size only rather than charge and size.

solvation. A state in which a molecule or ion is surrounded by solvent molecules.

solvation sphere. The shell of solvent molecules that surrounds an ion or solute.

special pair. A specialized pair of chlorophyll molecules in reaction centers that is the primary electron donor during the light-dependent reactions of photosynthesis.

specific heat. The amount of heat required to raise the temperature of 1 gram of a substance by 1°C.

specificity constant. See k_{cat}/K_m.

sphingolipid. An amphipathic lipid with a sphingosine (trans-4-sphingenine) backbone. Sphingolipids, which include sphingomyelins, cerebrosides, and gangliosides, are present in plant and animal membranes and are particularly abundant in the tissues of the central nervous system.

sphingomyelin. A sphingolipid that consists of phosphocholine attached to the C-1 hydroxyl group of a ceramide. Sphingomyelins are present in the plasma membranes of most mammalian cells and are a major component of myelin sheaths.

splice site. The conserved nucleotide sequence surrounding an exon-intron junction.

spliceosome. The large protein-RNA complex that catalyzes the removal of introns from mRNA precursors. The spliceosome is composed of small nuclear ribonucleoproteins.

splicing. The process of removing introns and joining exons to form a continuous RNA molecule.

sporulation. The response of certain organisms to unfavorable growth conditions. During sporulation, special cells, called spores, are formed. Spores remain dormant until favorable growth conditions are restored.

SRP. See signal recognition particle

SSB. See single-strand binding protein

stacking interactions. The weak noncovalent forces between adjacent bases or base pairs in single-stranded or double-stranded nucleic acids, respectively. Stacking interactions contribute to the helical shape of nucleic acids.

standard free-energy change ($\Delta G°'$). The free-energy change for a reaction under biochemical standard-state conditions.

standard reduction potential ($E°'$). A measure of the tendency of a substance to reduce other substances under biochemical standard-state conditions.

standard state. A set of reference conditions for a chemical reaction. In biochemistry, the standard state is defined as a temperature of 298K (25°C), a pressure of 1 atmosphere, a solute concentration of 1.0 M, and a pH of 7.0.

starch. A homopolymer of glucose residues that is a storage polysaccharide in plants. There are two forms of starch: amylose, an unbranched polymer of glucose residues joined by α-($1\rightarrow4$) linkages; and amylopectin, a branched polymer of glucose residues joined by α-($1\rightarrow4$) linkages with α-($1\rightarrow6$) linkages at branch points.

starvation. A period in which no food is digested or absorbed. During starvation, which begins about 16 hours after consumption of the last meal in humans, blood glucose originates from hepatic and renal gluconeogenesis, and fatty acids and proteins are catabolized for energy.

steady state. A state in which the rate of synthesis of a compound is equal to its rate of utilization or degradation.

stem-loop. See hairpin

stereoisomers. Compounds with the same molecular formula but different spatial arrangements of their atoms.

stereospecificity. The ability of an enzyme to recognize and act upon only a single stereoisomer of a substrate.

steroid. A lipid containing a fused, four-ring isoprenoid structure.

sterol. A steroid containing a hydroxyl group.

sticky end. An end of a double-stranded DNA molecule with a single-stranded extension of several nucleotides.

stomata. Structures on the surface of a leaf through which carbon dioxide diffuses directly into photosynthetic cells.

stop codon. See termination codon

strand invasion. The exchange of single strands of DNA from two nicked molecules having homologous nucleotide sequences.

stretch-gated ion channel. A membrane transport protein that allows ions to pass through the bilayer in response to changes in tension or turgor pressure on one side of the membrane.

stringent response. The production of phosphorylated guanylate compounds triggered by the presence of an uncharged tRNA molecule in the aminoacyl site of a ribosome during protein synthesis in prokaryotes. The net effect is a decrease in RNA synthesis when the cell is starved for amino acids.

stroma. The aqueous matrix of the chloroplast. The stroma is the site of the reductive pentose phosphate cycle, which leads to the reduction of carbon dioxide to carbohydrates.

stromal lamellae. Regions of the thylakoid membrane that are in contact with the stroma.

subcloning. Transferring cloned DNA between cloning vectors.

substrate. A reactant in a chemical reaction. In enzymatic reactions, substrates are specifically acted upon by enzymes, which catalyze the conversion of substrates to products.

substrate cycle. A pair of opposing, metabolically irreversible reactions that catalyzes a cycle between two pathway intermediates. Substrate cycles provide sensitive regulatory sites.

substrate-level phosphorylation. Phosphorylation of a nucleoside diphosphate by transfer of a phosphoryl group from a non-nucleotide substrate.

supercoil. A topological arrangement assumed by over- or underwound double-stranded DNA. Underwinding gives rise to negative supercoils; overwinding produces positive supercoils. Positively supercoiled DNA is not found in nature.

supersecondary structure. See motif

suppressor tRNA molecule. A mutant tRNA molecule containing an altered anticodon that permits it to bind to a termination codon. Such binding results in the incorporation of the amino acid carried by the tRNA into the growing polypeptide chain.

surfactant. See detergent

Svedberg unit (S). A unit of 10^{-13} seconds used for expressing the sedimentation coefficient, a measure of the rate at which a large molecule or particle sediments in an ultracentrifuge. Large S values usually indicate large masses.

symmetry-driven theory. See concerted theory of cooperativity and allosteric regulation

symport. The cotransport of two different species of ions or molecules in the same direction across a membrane by a transport protein.

synonymous codons. Different codons that specify the same amino acid.

synthase. A common name for an enzyme, often a lyase, that catalyzes a synthetic reaction.

synthetase. An enzyme that catalyzes the joining of two substrates and requires the input of the chemical potential energy of a nucleoside triphosphate. Synthetases are members of the IUB class of enzymes known as ligases.

T state. The less active conformation of an allosteric protein; opposite of R state.

TATA box. An A/T-rich DNA sequence found within the promoter of both prokaryotic and eukaryotic genes. In prokaryotes, the TATA box is located about 10 base pairs upstream of the transcription initiation site; in eukaryotes, it is located about 19 to 27 base pairs upstream of the transcription initiation site.

telomerase. A eukaryotic enzyme that extends the 3′ strand of DNA at the ends of chromosomes in order to prevent progressive shortening of the chromosome during successive rounds of replication.

telomere. The terminal region of eukaryotic chromosomes that consists of large numbers of 4–8 base pair repeats in which the nucleotides on one strand are predominantly deoxythymidylate and deoxyguanylate.

template.   A strand of DNA or RNA whose sequence of nucleotide residues guides the synthesis of a complementary strand.

template strand.   The strand of DNA within a gene whose nucleotide sequence is complementary to that of the transcribed RNA. During transcription, RNA polymerase binds to and moves along the template strand in the $3' \rightarrow 5'$ direction, catalyzing the synthesis of RNA in the $5' \rightarrow 3'$ direction.

termination codon.   A codon that is not normally recognized by any tRNA molecule but is bound by specific proteins that cause newly synthesized peptides to be released from the translation machinery. The three termination codons (UAG, UAA, and UGA) are also known as stop codons.

terpene.   One of a number of isoprenoid compounds found in plants.

tertiary structure.   The compacting of polymeric chains into one or more globular units within a macromolecule. In proteins, tertiary structure is stabilized mainly by hydrophobic interactions between side chains.

thermodynamics.   The branch of physical science that studies transformations of heat and energy.

thin-layer chromatography.   A chromatographic technique used to separate components of a mixture on a thin layer of porous material.

30 nm fiber.   A chromatin structure in which nucleosomes are coiled into a solenoid 30 nm in diameter.

−35 region.   A sequence found within the promoter of prokaryotic genes about 30–35 base pairs upstream of the transcription initiation site.

310 helix.   A secondary structure of proteins, consisting of a helix in which the carbonyl oxygen of each amino acid residue (residue n) forms a hydrogen bond with the amide hydrogen of the third residue further toward the C-terminus of the polypeptide chain (residue $n + 3$).

thylakoid lamella.   See thylakoid membrane

thylakoid membrane.   A highly folded, continuous membrane network suspended in the aqueous matrix of the chloroplast. The thylakoid membrane is the site of the light-dependent reactions of photosynthesis, which lead to the formation of NADPH and ATP. Also known as the thylakoid lamella.

tissue.   An association of cells that carry out the same functions.

T$_m$.   See melting point and phase-transition temperature

topoisomerase.   An enzyme that alters the supercoiling of a DNA molecule by cleaving a phosphodiester linkage in either one or both strands, rewinding the DNA, and resealing the break. Some topoisomerases are also known as DNA gyrases.

topology.   1. The arrangement of membrane-spanning segments and connecting loops in an integral membrane protein. 2. The overall morphology of a nucleic acid molecule.

TΨC arm.   The stem-and-loop structure in a tRNA molecule that contains the sequence thymidylate–pseudouridylate–cytidylate (TΨC).

trace element.   An element required in very small quantities by living organisms. Examples include copper, iron, and zinc.

transaminase.   An enzyme that catalyzes the transfer of an amino group from an α-amino acid to an α-keto acid. Transaminases require the coenzyme pyridoxal phosphate.

transcription.   The copying of biological information from a double-stranded DNA molecule to a single-stranded RNA molecule, catalyzed by a transcription complex consisting of RNA polymerase and associated factors.

transcription bubble.   A region of DNA that is unwound by RNA polymerase during transcription initiation.

transcription factor.   A protein that binds to the promoter region, to RNA polymerase, or to both during assembly of the transcription initiation complex. Some transcription factors remain bound during RNA chain elongation.

transcription initiation complex.   The complex of RNA polymerase and other factors that assembles at the promoter at the start of transcription.

transcriptional activator.   A regulatory DNA-binding protein that enhances the rate of transcription by increasing the activity of RNA polymerase.

transducer.   The component of a signal-transduction pathway that couples receptor-ligand binding with generation of a second messenger catalyzed by an effector enzyme.

transfection.   The introduction of foreign DNA into a cell via a virus or phage vector.

transfer ribonucleic acid (tRNA).   A class of RNA molecules that carry activated amino acids to the site of protein synthesis for incorporation into growing peptide chains. tRNA molecules contain an anticodon that recognizes a complementary codon in mRNA.

transferase.   An enzyme that catalyzes a group-transfer reaction. Transferases often require a coenzyme.

transformation.   See genetic transformation

transgenic organism.   An individual organism that carries recombinant DNA stably integrated in all of its cells.

transition.   A mutation resulting from a base alteration that occurs when a purine is substituted by the other purine, or when a pyrimidine is substituted by the other pyrimidine.

transition state.   An unstable, high-energy arrangement of atoms in which chemical bonds are being formed or broken. Transition states have structures between those of the substrates and the products of a reaction.

transition-state analog.   A compound that resembles a transition state. Transition-state analogs characteristically bind extremely tightly to the active sites of appropriate enzymes and thus act as potent inhibitors.

transition-state stabilization.   The increased binding of transition states to enzymes relative to the binding of substrates or products. Transition-state stabilization lowers the activation energy and thus contributes to catalysis.

translation.   The synthesis of a polypeptide whose sequence reflects the nucleotide sequence of an mRNA molecule. Amino acids are donated by activated tRNA molecules, and peptide bond synthesis is catalyzed by the translation complex, which includes the ribosome and other factors.

translation complex.   The complex of a ribosome and protein factors that carries out the translation of mRNA in vivo.

translation initiation complex.   The complex of ribosomal subunits, an mRNA template, an initiator tRNA molecule, and initiation factors that assembles at the start of protein synthesis.

translation initiation factor.   A protein involved in the formation of the initiation complex at the start of protein synthesis.

translational frameshifting.   The shift in reading frame that may occur during the translation of an mRNA molecule. The synthesis of some proteins requires translational frameshifting.

translocation.   1. The movement of the ribosome by one codon along an mRNA molecule. 2. The movement of a polypeptide through a membrane.

transport constant (K_{tr}). The substrate concentration at which the rate of transport across a membrane via a transport protein is half-maximal. K_{tr} is analogous to the Michaelis constant (K_m) of an enzyme.

transposable element. See transposon

transposon. A mobile genetic element that jumps between chromosomes or parts of a chromosome by taking advantage of recombination mechanisms. Also known as a transposable element or an integron.

transverse diffusion. The passage of lipid or protein molecules from one leaflet of a lipid bilayer to the other leaflet. Unlike lateral diffusion within one leaflet of a bilayer, transverse diffusion is extremely slow.

transversion. A mutation resulting from a base alteration that occurs when a purine is substituted by a pyrimidine, or when a pyrimidine is substituted by a purine.

triacylglycerol. A lipid containing three fatty acyl residues esterified to glycerol. Fats and oils are mixtures of triacylglycerols. Formerly known as a triglyceride.

tricarboxylic acid cycle. See citric acid cycle

triglyceride. See triacylglycerol

tRNA. See transfer ribonucleic acid

tumor promoter. A compound that greatly increases the development of tumors following exposure of an organism to a carcinogen.

turn. See loop

turnover. The dynamic metabolic steady state in which molecules are degraded and replaced by newly synthesized molecules.

turnover number. See catalytic constant

twist. The angle of rotation between adjacent residues within a helical macromolecule.

ubiquitin. A highly conserved protein in eukaryotic cells that is involved in protein degradation. The covalent attachment of one or more ubiquitin moieties to a protein targets that protein for intracellular hydrolysis.

uncompetitive inhibition. Inhibition of an enzyme-catalyzed reaction by an inhibitor that binds only to the enzyme-substrate complex, not to the free enzyme.

uncoupling agent. A compound that disrupts the usual tight coupling between electron transport and phosphorylation of ADP.

uniport. The transport of a single type of solute across a membrane by a transport protein.

unsaturated fatty acid. A fatty acid with at least one carbon-carbon double bond. An unsaturated fatty acid with only one carbon-carbon double bond is called a monounsaturated fatty acid. A fatty acid with two or more carbon-carbon double bonds is called a polyunsaturated fatty acid. In general, the double bonds of unsaturated fatty acids are of the cis configuration and are separated from each other by methylene ($-CH_2-$) groups.

urea cycle. A metabolic cycle consisting of four enzyme-catalyzed reactions that converts nitrogen from ammonia and aspartate to urea. Four ATP equivalents are consumed during formation of one molecule of urea.

v. See velocity

v_0. See initial velocity

vacuole. A fluid-filled organelle in plant cells that is a storage site for water, ions, or nutrients.

van der Waals force. A weak intermolecular force produced between neutral atoms by transient electrostatic interactions. Van der Waals attraction is strongest when atoms are separated by the sum of their van der Waals radii; strong van der Waals repulsion precludes closer approach.

van der Waals radius. The effective size of an atom. The distance between the nuclei of two nonbonded atoms at the point of maximal attraction is the sum of their van der Waals radii.

variable arm. The arm of a tRNA molecule that is located between the anticodon arm and the TΨC arm. The variable arm can range in length from about 3 to 21 nucleotides.

vector. See cloning vector

velocity (v). The rate of a chemical reaction, expressed as amount of product formed per unit time.

very low density lipoprotein (VLDL). A type of plasma lipoprotein that transports endogenous triacylglycerols, cholesterol, and cholesteryl esters from the liver to the tissues.

virus. A nucleic acid–protein complex that is capable of invading a host cell. A virus takes over the transcription and replication machinery of the host cell to replicate itself, using both its own and the host's gene products.

vitamin. An organic micronutrient that cannot be synthesized by an animal and must be obtained in the diet. Many coenzymes are derived from vitamins.

vitamin-derived coenzyme. A coenzyme synthesized from a vitamin.

VLDL. See very low density lipoprotein

Vmax. See maximum velocity

voltage-gated ion channel. A membrane transport protein that allows ions to pass through the bilayer in response to changes in the electrical properties of the membrane.

water-soluble vitamin. An organic micronutrient that is soluble in water. Specifically, water-soluble vitamins include the B vitamins and, in the case of primates and a few other organisms, ascorbic acid (vitamin C).

wax. A nonpolar ester that consists of a long-chain monohydroxylic alcohol and a long-chain fatty acid.

wobble position. The 5′ position of an anticodon, where non-Watson-Crick base pairing is permitted. The wobble position makes it possible for a tRNA molecule to recognize more than one codon.

X-ray crystallography. A technique used to determine secondary, tertiary, and quaternary structures of biological macromolecules. In X-ray crystallography, a crystal of the macromolecule is bombarded with X rays, which are diffracted and then detected electronically or on a film. The atomic structure is deduced by mathematical analysis of the diffraction pattern.

YAC. See yeast artificial chromosome

yeast artificial chromosome (YAC). A cloning vector that contains yeast centromeric DNA and a yeast origin of replication. A YAC can accommodate up to 500 kilobase pairs of insert DNA. The artificial chromosome can be introduced into a yeast cell, where it behaves like a normal chromosome during replication.

ylid. An organic molecule that has opposite ionic charges on adjacent covalently bonded atoms.

Z-DNA.   A conformation of oligonucleotide sequences containing alternating deoxycytidylate and deoxyguanylate residues. Z-DNA is a left-handed double helix containing approximately 12 base pairs per turn.

zero-order reaction.   A reaction whose rate is independent of reactant concentration.

zinc finger.   A structural motif often found in DNA-binding proteins. The finger is formed when a stretch of about 30 amino acids forms a loop whose base is anchored by Zn^{2+}. The Zn^{2+} is coordinated by two conserved histidine residues and two conserved cysteine residues at the base of the loop.

Z-scheme.   A zigzag scheme that illustrates the reduction potentials associated with electron flow through photosynthetic electron carriers.

zwitterion.   A molecule containing negatively and positively charged groups.

zymogen.   A catalytically inactive enzyme precursor that must be modified by limited proteolysis to become enzymatically active. Also known as a proenzyme.